Reviews of Physiology, Biochemistry and Pharmacology

For further volumes:
http://www.springer.com/series/112

Susan G. Amara · Ernst Bamberg ·
Bernd K. Fleischmann ·
Thomas Gudermann · Reinhard Jahn ·
William J. Lederer · Roland Lill ·
Bernd Nilius · Stefan Offermanns
Editors

Reviews of Physiology, Biochemistry and Pharmacology 161

Springer

Editors
Susan G. Amara
University of Pittsburgh
Pittsburgh, PA
USA
amara@pitt.edu

Ernst Bamberg
Max-Planck-Institut für Biophysik
Frankfurt
Germany
ernst.bamberg@mpibpfrankfurt.mpg.de

Bernd K. Fleischmann
Institut für Physiologie I
Life and Brain Center
Bonn
Germany
bernd.fleischmann@uni-bonn.de

Thomas Gudermann
Walther-Straub-Institut für Pharmakologie
und, Toxikologie
München
Germany
thomas.gudermann@lrz.uni-muenchen.de

Reinhard Jahn
Max-Planck-Institute for Biophysical
Chemistry
Göttingen
Germany
rjahn@gwdg.de

William J. Lederer
Medical Biotechnology Center
University of Maryland Biotechnology
Institute
Baltimore, MD
USA
jlederer@umaryland.edu

Roland Lill
University of Marburg
Medical Biotechnology Center
Marburg
Germany
lill@mailer.unimarburg.de

Bernd NiliusKatholieke
Universiteit Leuven
KUL
Campus Gasthuisberg
Herestraat 49
bus 802
Department of Molecular Cell Biology
Laboratory Ion Channel Research
B-3000 Leuven
Belgium
bernd.nilius@med.kuleuven.be

Stefan Offermanns
Max-Planck-Institut für Herz- und
Lungenforschung
Bad Nauheim
Germany
stefan.offermanns@mpi-bn.mpg.de

ISSN 0303-4240 e-ISSN: 1617-5786
ISBN 978-3-540-87628-1 e-ISBN 978-3-540-87631-1
DOI 10.1007/978-3-540-87631-1
Springer Heidelberg Dordrecht London New York

Printed on acid-free paper

Springer is part of Springer Science+Business Media (www.springer.com)

Contents

Regulation of Renin Release by Local and Systemic Factors

F. Schweda and A. Kurtz

Abstract The renin-angiotensin system (RAS) is critically involved in the regulation of the salt and volume status of the body and blood pressure. The activity of the RAS is controlled by the protease renin, which is released from the renal juxtaglomerular epithelioid cells into the circulation. Renin release is regulated in negative feedback-loops by blood pressure, salt intake, and angiotensin II. Moreover, sympathetic nerves and renal autacoids such as prostaglandins and nitric oxide stimulate renin secretion. Despite numerous studies there remained substantial gaps in the understanding of the control of renin release at the organ or cellular level. Some of these gaps have been closed in the last years by means of gene-targeted mice and advanced imaging and electrophysiological methods. In our review, we discuss these recent advances together with the relevant previous literature on the regulation of renin release.

Introduction

The renin-angiotensin system (RAS) regulates internal salt and fluid balance and blood pressure. Although the discovery of renin is credited to the work of Tigerstedt and Bergmann over a hundred years ago (Tigerstedt and Bergmann 1898), and although the RAS has been a focus of numerous studies ever since, substantial gaps remain in our understanding of how renin release is controlled.

According to the classical view of the RAS, the aspartyl protease renin is produced, stored, and released by the juxtaglomerular (JG) cells, which are located within the media layer of the afferent arterioles of the kidney (for reviews on renin release, see Friis et al. 2000, 2005, 2004; Hackenthal et al. 1990; Kurtz 1989; Kurtz and Wagner 1999; Schweda et al. 2007; Schweda and Kurtz 2004; Wagner and Kurtz 1998). Once secreted into circulation, renin cleaves the liver-derived angiotensinogen

F. Schweda (✉)
Institut für Physiologie, Universität Regensburg, 93040, Regensburg, Germany
e-mail: frank.schweda@klinik.uni-regensburg.de

at its N-terminus to form the decapeptide angiotensin I (Ang I). Ang I is rapidly converted into angiotensin II (Ang II) by the cleavage of two amino acids, a reaction catalyzed by the angiotensin-converting enzyme (ACE). Ang II is considered the main biological effector of the RAS as it mediates its classical functions, such as vasoconstriction, stimulation of renal salt reabsorption, enhancement of catecholamine and aldosterone release, and stimulation of thirst and salt appetite. Since all of the aforementioned effects of Ang II increase blood pressure, the effect that led to the discovery of renin by Tigerstedt and Bergman (Tigerstedt and Bergmann 1898), blood pressure stabilization is the main physiological function of the RAS. This straight forward picture of the RAS has since become more complicated with the discovery of renin expression in extrarenal tissues such as the heart, blood vessels, adrenal glands, reproductive organs, pancreas, brain, and mast cells (Dzau et al. 1987; Ekker et al. 1989; Ganten et al. 1971; Itskovitz et al. 1992; Pandey et al. 1984; Paul et al. 1988, 1993; Mackins et al. 2006; Veerappan et al. 2008). Even renin in the kidney is found not only in its classical position at the vascular pole of the glomerulus but also in the proximal tubules, the connecting tubule, and the collecting ducts (Chen et al. 1994; Moe et al. 1993; Prieto-Carrasquero et al. 2004, 2005; Rohrwasser et al. 2003, 1999; Taugner et al. 1982). In addition to renin, the other components of the renin-angiotensin cascade are more or less completely present in these aforementioned tissues, where they form locally active tissue renin-angiotensin systems that are involved in the regulation of organ function in physiological and pathophysiological states. (For reviews of local RAS, please see Bader et al. 2001; Baltatu and Bader 2003; Danser 2003; Dzau 1993; Ichihara et al. 2004; Kobori et al. 2007; Leung 2007; Leung and Sernia 2003; Paul et al. 2006; Re 2004).

The classical view of the systemic RAS, in which Ang II is the only biological effector, has been complicated by the discovery of additional bioactive components of the RAS. For instance, Ang II can be further metabolized by aminopeptidases to angiotensin III (Ang III; 7 amino acids) and angiotensin IV (Ang IV; 6 amino acids) (Reudelhuber 2005). Furthermore, a biologically active angiotensin peptide consisting of seven amino acids, Ang(1-7), can be formed from Ang I or from Ang II by the angiotensin-converting enzyme ACE2 (Crackower et al. 2002; Tipnis et al. 2000). Finally, renin and its precursor prorenin can bind to the renin/prorenin receptor, thereby activating signaling cascades independent of their enzymatic activity (Danser et al. 2007; Nguyen 2006). These discoveries about the local RAS and the biochemistry of angiotensin peptides have been the focus of a number of recent review articles (Kobori et al. 2007; Paul et al. 2006; Reudelhuber 2005) and are not within the scope of the present review. Instead, we focus on recent advances in the understanding of the control of renin release from the renin-producing JG cells.

The Process of Renin secretion

Renin is encoded by a single gene in humans and in most animal species, but not in mice. Laboratory mice, widely used in renin research, fall into two categories: strains (e.g., C57BL/6 and BALB/c) that carry one renin gene, designated Ren-1c, and other strains (e.g., 129Sv and Swiss) that have two renin genes, Ren-1d and Ren-2. The latter two genes arose through gene duplication and are therefore located in close proximity to each other on chromosome 1 (Abel and Gross 1990; Dickinson et al. 1984). All three renin genes encode highly homologous proteins (97% homology at the amino acid level), which nevertheless possess different glycosylation potentials (Sigmund and Gross 1991). While the Ren-1 proteins can be glycosylated at three asparagine residues, the Ren-2 enzyme lacks these putative glycosylation sites.

The renin gene in rats and mice consists of nine exons and eight introns, while the human renin gene contains an additional mini-exon (exon Va) encoding only three amino acids (Hardman et al. 1984). The gene is transcribed, and the mRNA is processed and finally translated into the protein preprorenin (401 amino acids). Preprorenin is translocated into the endoplasmic reticulum (ER), where a 20-residue ER-targeting signal sequence is cleaved from preprorenin, thereby generating prorenin (Fig. 1). The enzymatically inactive prorenin can either be released continuously

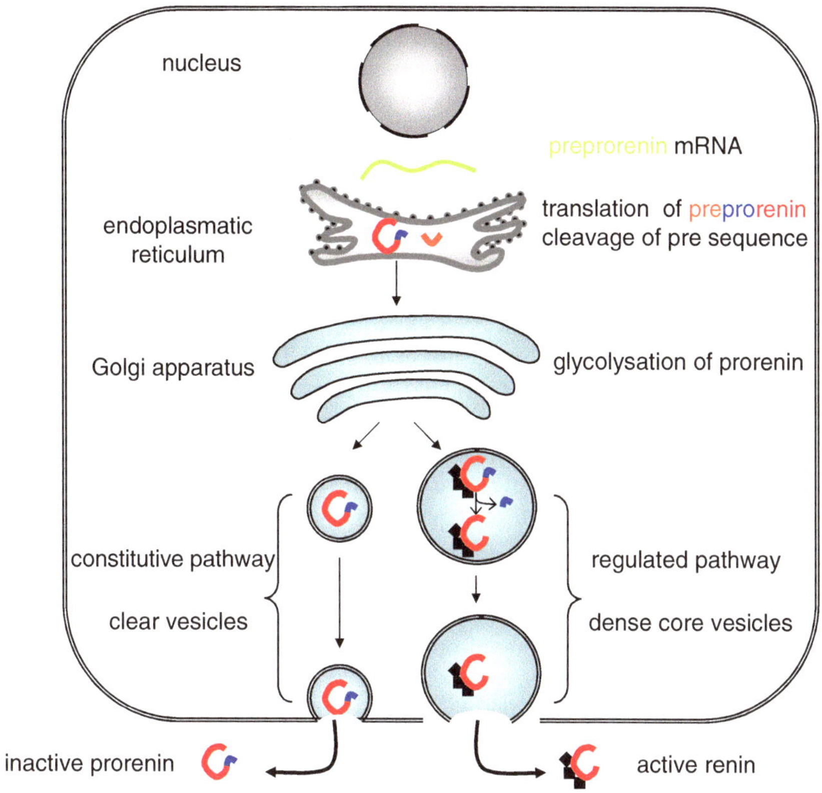

Fig. 1 Intracellular synthesis of renin in juxtaglomerular cells. After transcription, preprorenin is transferred into the endoplasmic reticulum, where the presequence is cleaved. Prorenin is transported to the Golgi apparatus, where it can be glycosylated and tagged for the regulated exocytosis pathway. Prorenin is activated by cleavage of its prosequence in the dense core vesicles destined for regulated exocytosis. Untagged prorenin that is sorted for constitutive exocytosis remains inactive, since its prosequence is not cleaved

into circulation or be packed into secretory vesicles by the Golgi apparatus, leading to the formation of rhomboid or fusiform protogranules. During the maturation of protogranules to mature granules, prorenin is further processed into active renin via cleavage of the 46-residue prosequence from its N-terminus (Fig. 1).While the constitutive pathway of prorenin release is determined primarily by the renin synthesis rate, reflected in the abundance of renin mRNA, the release of active renin from the secretory granules is rapidly regulated. In this way, stimulation or suppression of the renin system in the long term (over days or weeks) regulates plasma levels of both active renin and prorenin, while in the short term (up to 2 h) only active renin levels are changed (Toffelmire et al. 1989).

Similar to other aspartyl proteases, renin has a bilobular form and its catalytic site is buried in the cleft between the two lobes (Blundell et al. 1983). Since the prosegment hinders the access of angiotensinogen to the active site, prorenin was originally thought to be biologically inactive. However, as mentioned above, this view is questioned since the discovery of the renin/prorenin receptor.

A critical factor determining whether prorenin is sorted into the pathway of constitutive or regulated release appears to be the glycosylation of prorenin with mannose-6-phosphate residues. The importance of the glycosylation of prorenin for storage and for generation of secretory granules has recently been suggested in studies using renin knockout (KO) mice. As mentioned above, the Ren-2 protein lacks the potential glycosylation sites that are present in the Ren-1 proteins. Genetic deletion of the Ren-1d gene in a mouse strain with two renin genes resulted in a complete loss of renin-containing granules in the JG cells (Clark et al. 1997), whereas the granules are intact in Ren-2 KO mice (Sharp et al. 1996).

The proteolytic mechanism by which prorenin is cleaved from its prosequence inside the protogranules has not been identified. Since cathepsin B coexists with renin in the secretory granules (Matsuba et al. 1989; Taugner and Hackenthal 1988) and has the potential to convert prorenin to renin (Jutras and Reudelhuber 1999; Neves et al. 1996; Wang et al. 1991), this protease is a putative candidate for processing prorenin. Kallikreins have also been suggested as possible activators of prorenin (Kikkawa et al. 1998; Yokosawa et al. 1979). In addition, the proprotein convertase PC5 was found to be capable of activating prorenin in cell culture experiments (Laframboise et al. 1997; Mercure et al. 1996). However, despite low expression of PC5 mRNA in the kidney, attempts to demonstrate the presence of PC5 protein in the human kidney by immunohistochemistry have failed (Mercure et al. 1996). Since the activation of prorenin is obviously an important step in the biological activity of the RAS, further studies should focus on identifying the protease responsible. Moreover, the central position of this activation step in the renin-angiotensin cascade leads to the speculation that this proteolytic reaction may itself be regulated such that the activity of the RAS is controlled not only by the rate of renin release but also by the transformation rate of inactive prorenin into active renin within the secretory granules.

The mode of renin release from JG cells has been a matter of debate for a long time. However, despite reports arguing for different mechanisms of renin extrusion (King et al. 1993), convincing morphological evidence has accumulated in favor of an exocytotic release of renin by JG cells (Ogawa et al. 1995; Taugner et al. 1984a). In a recent study using multiphoton fluorescence imaging, renin vesicles that were located deep in the renin-producing cells disappeared without showing any movement (Peti-Peterdi et al. 2004), suggesting that the granules do not necessarily have to move toward the cell surface to contact the plasma membrane. This somewhat unexpected finding may be explained by channel-like invaginations in the plasma membrane of JG cells that had been observed earlier (Peter 1976; Ryan et al. 1982) and that likely result from the preceding fusion of vesicles with the plasma membrane.

Functional evidence for renin exocytosis was derived from the in vitro observation that renin is released from afferent arterioles not continuously, but instead in an episodic,"quanta-like" fashion (Skott 1986). In line with these data, acute stimulation of renin secretion in vivo markedly reduces the number of renin storage granules, while the average size of the remaining granules remains constant (Rasch et al. 1998). It can be inferred from these findings that upon stimulation, a renin storage granule releases its contents in an "all or none" fashion typical of exocytosis. This interpretation has been corroborated by direct visualization of renin release (Peti-Peterdi et al. 2004). Finally, electrophysiological evidence for the exocytosis of renin also exists,which is based on the rationale that in the process of exocytosis the membrane of the renin storage granule fuses with the plasma membrane, thereby enlarging the cell surface of the JG cell. Since whole-cell electrical capacitance changes in parallel with the cell surface, exocytotic events are accompanied by increases in whole-cell capacitance (Neher and Marty 1982). Applying this concept to isolated mouse JG cells, patch-clamp studies have demonstrated that cell membrane capacitance is increased by classical stimulators of renin release such as cAMP, which argues for membrane insertion via exocytosis (Friis et al. 1999).

How the fusion of the renin storage granule with the plasma membrane occurs and how it is regulated in JG cells is not understood. In other secretory cells the exocytotic process involves a sequence of vesicle docking, tethering, and finally fusion of the vesicle membrane with the plasma membrane. Three SNARE proteins are involved in these processes (Weber et al. 1998). Syntaxin and SNAP25 are localized at the plasma membrane (target or t-SNAREs), while synaptobrevin resides at the vesicle membrane (vesicle or v-SNARE) (Weber et al. 1998). In addition to the SNARE proteins, at least two other protein families, the Rab proteins and the SM proteins (Sec1/Munc18-like proteins), are required for vesicle fusion (Jahn et al. 2003). Since a broad range of cell types and organisms appear to use the above-mentioned processes to achieve vesiclefusion, it might be speculated that the same exocytosis machinery exists in JG cells. Thus far, however, no data have been reported confirming this hypothesis.

Intracellular signaling pathways controlling renin release

Similar to hormone release from other secretory cells, renin secretion is controlled by the classical intracellular second messengers cyclic AMP, cyclic GMP, and calcium (Ca^{2+}) (Fig. 2).

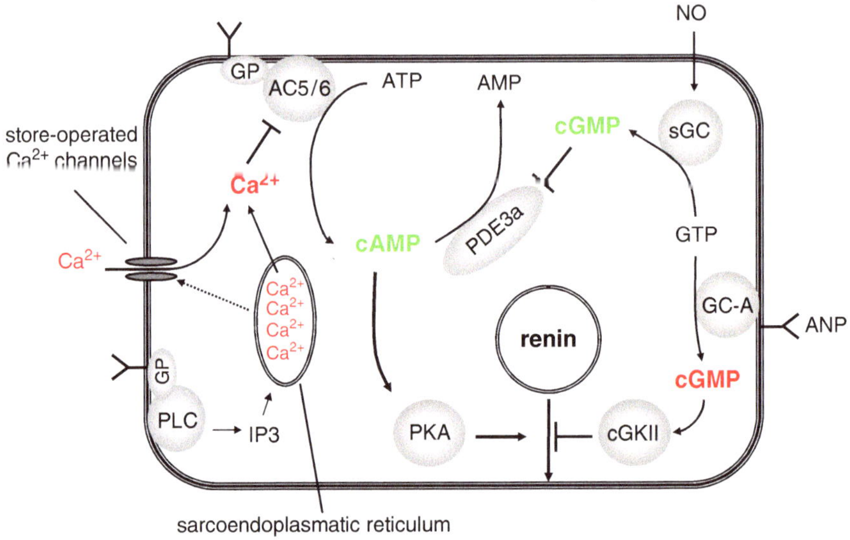

Fig. 2 Cellular control of renin release. Renin release from juxtaglomerular cells is regulated by cyclic AMP(*cAMP*), calcium (*Ca^{2+}*), and cyclic GMP (*cGMP*). For further explanation see the section entitled "Intracellular signaling pathways controlling renin release"

Cyclic AMP

Among the second messengers regulating renin release, cAMP plays a central stimulatory role. This is suggested by the fact that all hormones stimulating renin release exert their effects via elevations of intracellular cAMP levels. For instance, the prostaglandins E_2 and I_2 (prostacyclin) (Jensen et al. 1996), dopamine (Kurtz et al. 1988b), calcitonin gene-related peptide (Kurtz et al. 1988c), pituitary adenylyl cyclase-activating polypeptide (PACAP) (Hautmann et al. 2007), and adrenomedullin (Jensen et al. 1997) directly stimulate renin release from isolated JG cells and increase intracellular cAMP concentrations. Furthermore, the sympathetic nervous system, representing a very effective and physiologically important regulator of renin release, stimulates renin secretion via activation of the cAMP pathway through β-adrenoreceptors. Pharmacological stimulation of β-adrenoreceptors enhances renin release in a broad range of experimental models including single JG cells and in vivo experiments (Baumbach and Skott 1986; Friis et al. 1999; Keeton and Campbell 1980; Kurtz et al. 1984; Vandongen et al. 1973; Weinberger et al. 1975). Moreover, β-adrenoreceptors mediate the increase in renin release in response to renal nerve stimulation (Holdaas et al. 1981). Consistent with the central role of β-adrenoreceptors, β_1/β_2-adrenoreceptor KO mice show a low plasma renin concentration (PRC) both at baseline and in response to a variety of physiological challenges of the renin system (Kim et al. 2007a).

Besides receptor-mediated activation, direct activation of the adenylyl cyclases (ACs) by forskolin strongly increases renin release in isolated JG cells (Grünberger et al. 2006; Kurtz et al. 1984), as well as in the renin-producing cell line As4.1

(Grünberger et al. 2006; Klar et al. 2002). Finally, membrane-permeable cAMP analogs or direct intracellular applications of cAMP via a patch pipette rapidly stimulate renin secretion (Friis et al. 1999), indicating that the effect of cAMP is in fact related to its cytosolic level.

The fundamental physiological importance of the cAMP pathway has been emphasized recently in mice carrying a conditional deletion of the stimulatory G protein $G_s\alpha$ in JG cells (Chen et al. 2007). Since $G_s\alpha$ mediates the activation of adenylyl cyclases by G protein-coupled receptors, such as the β-adrenoreceptor, the PGE_2 receptors EP2 and EP4, and the PGI_2 receptor IP, deletion of $G_s\alpha$ disrupts the cAMP formation normally induced by these stimulators of renin release. In fact, $G_s\alpha$ KO mice show a marked reduction in PRC and renin gene expression (Chen et al. 2007). Moreover, the well-known stimulations of PRC by catecholamines, by a drop in blood pressure, or by activation of the macula densa mechanism, were virtually absent in $G_s\alpha$ KO mice, indicating that these central control mechanisms of renin release depend on the generation of cAMP by JG cells (Chen et al. 2007).

The cytosolic concentration of cAMP is determined by the balance of cAMP formation by adenylate cyclases and by the rate of cAMP hydrolysis to 5'-AMP through the action of cAMP phosphodiesterases (PDEs). In this way, nonselective inhibition of PDE activity by xanthine derivatives, such as isobutyl methylxanthine (IBMX) or theophylline, stimulates renin release. Pharmacological studies using selective blockers of different PDE isoforms suggest that it is primarily the activity of isoforms PDE-3 and PDE-4 that determines the intracellular levels of cAMP in JG cells (Chiu et al. 1996, 1999; Chiuand Reid 1996). In fact, a strong expression of these enzymes was confirmed in single JG cells and in the JG cell line As4.1 (Friis et al. 2002; Klar et al. 2002). In line with this finding, selective inhibition of PDE-3 and of PDE-4 strongly stimulates renin secretion in vivo and in vitro. PDE-3 activity has been shown to be regulated by cGMP, while this has not yet been reported for PDE-4 activity. An increase in cytosolic cGMP inhibits PDE-3, thus elevating cAMP levels (Beavo 1995). Through this pathway, cGMP indirectly stimulates renin secretion, which is of major physiological importance for the influence of nitric oxide (NO) on renin release (see below).

Despite the central role of cAMP in the stimulation of renin release, knowledge of the mechanisms by which cAMP triggers exocytosis of renin storage vesicles is largely limited to the observation that this involves a PKA-dependent step (Castrop et al. 2005; Friis et al. 2002; Kurtz et al. 1998a). The downstream signaling is currently unclear; it will be a future challenge to identify the targets of PKA that mediate the stimulation of renin exocytosis.

Cyclic GMP

In contrast to cAMP, which stimulates renin release, cGMP can both stimulate and inhibit renin secretion. This dual action of cGMP on renin secretion has led to controversy about the role of cGMP in the control of renin secretion. The controversy mainly centered around studies investigating the effects of endogenous factors that elevate cGMP levels in JG cells, such as NO or atrial natriuretic peptide (ANP). While it is now widely accepted that NO can stimulate renin release in vivo and in vitro, contradictory results have also been reported (Kurtz and Wagner 1998). The same holds true for the effects of ANP, for which both stimulation and inhibition of renin release have been described (Kurtz and Wagner 1998).

This confusing scatter of results becomes clearer when comparing the effects of a direct application of cGMP with the effects of membrane-permeable cGMP analogs. When applied in high concentrations, these cGMP analogs frequently inhibit renin release from isolated perfused kidneys, kidney slices, and primary cultures of JG cells, as well as from dispersed JG cells (Greenberg et al. 1995; Henrich et al. 1988; Kurtz et al. 1986b, 1998a; Noble et al. 1994; Schricker and Kurtz 1993). In contrast, direct injection of unmodified cGMP into JG cells with a patch pipette significantly increases renin exocytosis (Friis et al. 2002). Apparently, cGMP and stable cGMP analogs do not exert identical effects on renin secretion, which is best explained by the different affinity of native cGMP and stable cGMP analogs to intracellular target molecules. Thus, cAMP-degrading phosphodiesterases such as PDE-3 are inhibited more efficiently by cGMP than by stable cGMP analogs, while the latter show a much higher affinity for cGMP-activated kinases (cGKs) (Butt et al. 1992). Moreover, since membrane-permeable cGMP analogs are resistant to intracellular degradation, these agents presumably circumvent the compartmentalization of cGMP signaling that appears to be necessary to segregate the cellular responses to stimulation with ANP or NO. Such an compartmentalization has been reported in other cell types, though not yet in JG cells (Piggott et al. 2006).

As mentioned above, PDEs, in particular PDE-3 and PDE-4, play a major role in adjusting cytosolic cAMP levels in JG cells. That PDE-3 plays a mediator function was recognized when the mechanisms by which NO influences renin secretion were clarified. It was found that pharmacological inhibition of PDE-3 stimulates renin secretion in vivo and in vitro, and that PDE-3 inhibition largely mimics the effects of NO on renin secretion (Chiu and Reid 1996; Chiu et al. 1999; Friis et al. 2002; Kurtz et al. 1998b). Moreover, NO donors increase intracellular cAMP concentration and renin secretion in primary cultures of JG cells, similar to the effects of PDE-3 inhibitors (Kurtz et al. 1998a). This interplay between the cGMP and cAMP pathways has been confirmed in vivo (Beierwaltes 2006), in isolated perfused kidneys (Kurtz et al. 1998b), and in patch-clamp experiments on single JG cells, in which cGMP stimulated renin exocytosis in a PDE-3-, cAMP-, and PKA-dependent manner (Friis et al. 2002).

JG cells express cGMP-dependent protein kinase I (cGKI) and cGMP-dependent protein kinase II (cGKII). While cGKII occurs mainly in association with renin storage granules, cGKI is localized in the cytosol, suggesting a functional role in the regulation of exocytosis (Gambaryan et al. 1996, 1998). In isolated perfused rat kidneys nonselective pharmacological stimulation of cGKs inhibits renin secretion (Gambaryan et al. 1998; Wagner et al. 1998). Similarly, in microdissected glomeruli with attached afferent arterioles stable cGMP analogs reduce the stimulation of renin release by cAMP, and this inhibitory effect of cGMP is completely abolished by blockade of cGKs (Gambaryan et al. 1998). Finally, stable cGMP analogs suppress renin release in primary cultures of JG cells isolated from wild-type or cGKI-deficient kidneys, but not from cGKII-deficient kidneys (Gambaryan et al. 1998; Wagner et al. 1998). It can be concluded from these studies that cGKII mediates the suppression of renin release due to cGMP. How activation of cGKII inhibits the exocytosis of renin storage granules remains to be clarified.

The factors determining whether the stimulatory or the inhibitory effect of endogenous cGMP on renin secretion predominates are poorly understood. Since PDE-3 has a higher affinity for cGMP than cGKII, and since the inhibitory effect of cGKII activation can in principle attenuate the stimulatory effect of cAMP on renin secretion, it is reasonable to speculate that renin secretion is stimulated by more moderate increases in cGMP level but inhibited by strong increases in cGMP level (Kurtz and Wagner 1998). Moreover, the intracellular locations of cGMP-generating enzymes and cGMP target molecules suggest that intracellular compartmentalization is a major factor defining the overall effect of cGMP on renin secretion. Soluble guanylate cyclase (sGC), which is activated by NO, is a cytosolic enzyme like PDE-3; cGMP derived from sGC may therefore predominantly stimulate renin secretion, via the cAMP pathway. Membrane-bound particulate guanylate cyclase (pGC), which is activated by ANP, is found in close proximity to vesicles associated with cGKII. This suggests that cGMP derived from pGC predominantly inhibits renin secretion via cGKII. Although such a spatial segregation of cGMP signals induced by ANP and NO has recently been demonstrated in both HEK cells and vascular smooth muscle cells (Nausch et al. 2008; Piggott et al. 2006), no data supporting this hypothesis in JG cells have yet been published.

Calcium

While cAMP is the primary stimulator of renin release, free cytosolic Ca^{2+} concentration is considered the primary inhibitor of renin release. In the vast majority of secretory cells, an increase in the intracellular Ca^{2+} concentration initiates and supports exocytosis. Therefore, the inverse relationship between Ca^{2+} and renin secretion in JG cells, that besides in the JG cells has only been shown in parathyroid gland cells (Cohen et al. 1997), has been called the "calcium paradox" of renin release. Since the effects of Ca^{2+} on renin release have recently been reviewed (Schweda and Kurtz 2004), the present article focuses on new advances in our understanding of the inhibition of renin exocytosis by intracellular Ca^{2+} concentration.

The present knowledge about the effects of intracellular Ca^{2+} concentration on renin release is based on numerous studies using indirect approaches to manipulate the intracellular Ca^{2+} concentration. Thus, classical vasoconstrictors such as Ang II, endothelin, vasopressin, and norepinephrine increase Ca^{2+} influx or intracellular Ca^{2+} levels in JG cells (Grünberger et al. 2006; Ichihara et al. 1995; Kurtz et al. 1984, 1986c) and inhibit renin release in a Ca^{2+}-dependent fashion (Moe et al. 1991; Takagi et al. 1988b). Also, receptor-independent measures that modulate the cytosolic Ca^{2+} concentration produce inverse changes of renin release. For instance, blocking the sarcoendoplasmic Ca^{2+}-ATPase with thapsigargin induces store-operated Ca^{2+} influx and elevates the intracellular Ca^{2+} concentration in JG cells, which suppresses renin release from isolated perfused kidneys and cultured JG cells (Grünberger et al. 2006; Schweda et al. 2000; Yao et al. 2003). On the other hand, reducing intracellular Ca^{2+} levels with the intracellular Ca^{2+} chelator BAPTA stimulates renin release (Moe et al. 1991; Ortiz-Capisano et al. 2007b, 2007c).

The influence of extracellular Ca^{2+} concentration on renin release has also been the focus of several studies. Lowering the extracellular Ca^{2+} concentration stimulates renin secretion from isolated perfused kidneys, isolated glomeruli, kidney slices, renal cortical cell suspensions, and cultured JG cells (Hackenthal et al. 1990; Kurtz and Wagner 1999; Taugner et al. 1988), while in vivo Ca^{2+} infusions reduce renin secretion in nonfiltering kidneys of dogs (Watkins et al. 1976). Since the extracellular and intracellular Ca^{2+} concentrations correlate positively with each other in JG cells (Kurtz and Penner 1989), the inverse relationship between renin release and extracellular Ca^{2+} was assumed to result from parallel changes in the intracellular Ca^{2+} concentration. In addition to this possibility, the Ca^{2+}-sensing receptor may participate in the regulation of renin release by extracellular Ca^{2+} concentration (Ortiz-Capisano et al. 2007a).

The unusual effect of cytosolic Ca^{2+} on renin exocytosis has been the starting point for numerous studies attempting to identify the intracellular mechanisms underlying this effect. In fact, at least three intracellular pathways have been proposed to be involved in the inhibitory effects of Ca^{2+}: Ca^{2+}/calmodulin-dependent processes, such as activation of myosin light chain kinase; activation of protein kinase C; and modulation of ion channel activities in the plasma membrane or in the membrane of the renin-containing granule (reviewed in Hackenthal et al. 1990; Kurtz 1989; Schweda and Kurtz 2004). This list of downstream targets of Ca^{2+} must be expanded to include the Ca^{2+}-inhibited adenylyl cyclases (ACs) (Grünberger et al. 2006; Ortiz-Capisano et al. 2007c). Of the nine membrane-bound ACs, AC5 and AC6 are inhibited by physiological increases in cytosolic Ca^{2+}. Thus, in cells expressing AC5 and AC6, an increase in cytosolic Ca^{2+} can inhibit cAMP generation and reduce intracellular cAMP levels, which correspond to the primary intracellular stimulator of renin release. In fact, an inverse relationship between the intracellular Ca^{2+} concentration and cAMP levels has been demonstrated in JG cells. On the one hand, application of the intracellular Ca^{2+} chelator BAPTA-AM elevates intracellular cAMP levels and stimulates renin release from JG cells, most likely by lowering the intracellular concentration of free Ca^{2+} (Ortiz-Capisano et al. 2007b, 2007c). In addition, increases in intracellular Ca^{2+} concentration due to endothelin-1, Ang II, or thapsigargin occur in parallel with reductions in intracellular cAMP levels and inhibition of renin release (Grünberger et al. 2006). The Ca^{2+} liberators do not suppress renin release when intracellular cAMP levels are clamped, and chelation of intracellular Ca^{2+} does not enhance cAMP levels or renin release following blockade of AC activity (Grünberger et al. 2006; Ortiz-Capisano et al. 2007c). Therefore, these data suggest that the inverse relationship between renin release and the cytosolic Ca^{2+} concentration depends, in fact, on changes in intracellular cAMP levels. The functional role of AC inhibition by Ca^{2+} has been further corroborated in gene knockdown studies (Grünberger et al. 2006) and by pharmacological studies using the drug NKY-80, which shows selective binding to AC5 in preference to the non-calcium-inhibited isoforms AC2 and AC3 (Onda et al. 2001; Ortiz-Capisano et al. 2007b). Consistent with the functional data, expression of AC5 and AC6 mRNA has been found in JG cells, but protein expression has so far been demonstrated only for AC5 (Grünberger et al. 2006; Ortiz-Capisano et al. 2007b; Wang and Brown 2004).

Taken together, these data could provide an elegant explanation for the Ca^{2+} paradox, according to which the primary stimulatory cAMP pathway is inversely related to intracellular Ca^{2+} concentration via Ca^{2+}-induced inhibition of ACs. It is important to point out that this explanation does not exclude the possibility that other mechanisms contribute to the Ca^{2+}-dependent inhibition of renin release at points further upstream or downstream of cAMP effects. As outlined above, cAMP is a critical factor in stimulating renin release. It is therefore reasonable to assume that downregulation or pharmacological blockade of ACs, as in the above-mentioned studies, can suppress and mask the effects of other mechanisms regulating renin release in a Ca^{2+}-dependent fashion. Further research using more complex models and in vivo experiments is needed to verify the physiological role of AC5 and AC6 in regulating renin release.

Electrophysiology of JG cells

Since electrical events participate in the regulation of exocytosis in a variety of tissues, the electrical properties of JG cells have been studied in some detail. Most of these studies have been performed on JG cells in situ in afferent arterioles, and they have reported a strong electrical coupling of JG cells to the neighboring vascular smooth muscle cells (Bührle et al. 1985; Kurtz and Penner 1989; Russ et al. 1999). Some of the results obtained in these studies have been confirmed and extended in isolated single JG cells.

There is direct electrophysiological evidence for at least three potassium (K^+) conductances in JG cells: The first conductance is associated with a delayed outwardly rectifying K^+ current (Friis et al. 1999, 2003; Kurtz and Penner 1989) that has been identified as the BK_{Ca} ZERO splice variant (Friis et al. 2003); the second conductance is associated with the inwardly rectifying K^+ channels Kir (Friis et al. 1999; Kurtz and Penner 1989; Leichtle et al. 2004); the third conductance is associated with K_{ATP} channels (Russ et al. 1999). Moreover, there is good evidence for the existence of Ca^{2+}-activated Cl^- channels and the $Na^+/K^+/2Cl^-$ cotransporter (NKCC1) in the plasma membrane of JG cells (Castrop et al. 2005; Kaplan et al. 1996; Kurtz and Penner 1989).

Because of the marked inhibitory effect of the cytosolic Ca^{2+} concentration on renin release, several studies have examined the role of Ca^{2+} channels in JG cells. While early evidence suggested the existence of potential-operated Ca^{2+} channels in JG cells (Churchill 1985), this evidence has not been confirmed by further functional studies or by direct measurements of currents or Ca^{2+} concentration (Kurtz et al. 1990; Scholz and Kurtz 1995) (for detailed discussion, see Schweda and Kurtz 2004). A recent study demonstrating mRNA and protein expression of L-type Ca^{2+} channels in JG cells found that these channels are activated at very positive membrane potentials (Friis et al. 2003). Since such depolarization will hardly be achieved in JG cells under physiological conditions, these data may explain why several studies have failed to detect a direct functional role of L-type Ca^{2+} channels in the regulation of renin release (Schweda and Kurtz 2004). Finally, there is good, albeit indirect, evidence for store-operated Ca^{2+} entry (SOC) triggered by the filling state of the internal $Ca2^{+}$ stores (Grünberger et al. 2006; Kurtz and Penner 1989; Yao et al. 2003). The identity of the channels underlying this Ca^{2+} influx is currently unclear. It will be interesting to investigate the role of ORAI proteins, as well as TRPC channels, in the regulation of Ca^{2+} influx and renin release in JG cells, since these channels contribute to or mediate SOC in other cells (Smyth et al. 2006).

The resting membrane potential of JG cells in isolated, nonpressurized afferent arterioles is between –60 and –80 mV (Bührle et al. 1985; Kurtz and Penner 1989; Russ et al. 1999), and probably increases to approximately -40 mV in pressurized arterioles in situ (Loutzenhiser et al. 1997). It was an early observation that vasoconstrictors depolarize JG cells in situ (Bührle et al. 1985, 1986). This depolarization probably results from inhibition of Kir by certain vasoconstrictors such as Ang II via a G protein-mediated process (Kurtz and Penner 1989). A second mechanism likely to contribute to membrane depolarization by vasoconstrictors is related to their Ca^{2+} mobilizing activity, which leads to an activation of Ca^{2+} activated Cl^{-} channels, Cl^{-} efflux, and membrane depolarization. Since vasoconstrictor hormones uniformly depolarize JG cells and at the same time inhibit renin secretion, it has been speculated that the membrane potential is an important determinant for exocytosis, such that depolarization inhibits renin secretion. This assumption has been supported by a variety of studies showing that renin secretion is attenuated by depolarization induced either by an increase in extracellular K^{+} concentration or by inhibition of K^{+} channels (Churchill 1980; Ginesi et al. 1983; Kurtz et al. 2000, 1990). On the other hand, membrane hyperpolarization induced by K^{+} channel activation or Cl^{-} channel inhibition facilitates renin secretion (Ferrier et al. 1989; Jensen et al. 1998; Jensen and Skott 1996; Nabel et al. 1999). Consistent with this reasoning, pharmacological blockade of $Na^{+}/K^{+}/2Cl^{-}$ cotransport activity with furosemide led to the hyperpolarization of single JG cells and stimulated renin exocytosis at the same time (Castrop et al. 2005). Both effects of furosemide appeared to depend on the NKCC1 isoform, since they were not observed in NKCC1 KO mice (Castrop et al. 2005). Since Ca^{2+} inhibits renin secretion, it would be natural to speculate that potential-operated Ca^{2+} channels mediate the influence of membrane potential on renin secretion. However, as mentioned above, these channels are activated only at very high membrane potentials (Friis et al. 2003), which calls into question a functional role for them under physiological conditions. Therefore, further research is needed to investigate other mechanisms by which the membrane potential regulates renin exocytosis.

Role of intercellular communication for renin secretion

Renin-producing JG cells are intensely coupled to each other, as well as to the neighboring mesangial and endothelial cells by gap junctions (Taugner et al. 1984a, 1984b, 1978). Gap junctions are formed by connexins, which compose a group of at least 20 members and are named according to their molecular mass in kilodaltons (Söhl and Willecke 2004). Six connexins assemble to form a hemichannel called a connexon. Connexons from one cell can combine with a connexon on an adjacent cell to form a complete cell-cell channel (gap junction), thereby enabling intercellular communication (Söhl and Willecke 2004). Alternatively, a connexon can serve as a single transmembrane channel that, for instance, enables the release of intracellular ATP into the extracellular compartment, thereby mediating intercellular communication in a paracrine way (Evans et al. 2006; Saez et al. 2003).

Because of the intense gap junctional coupling in the juxtaglomerular apparatus and the renal vasculature, several studies have examined the role of connexins in the regulation of renal hemodynamics (recently reviewed in Wagner 2008). Moreover, gap junctions are involved in the regulation of renin release. Thus, pharmacological blockade of gap junctions by 18α-glycyrrhetinic acid abrogated the pressure-dependent regulation of renin release in isolated perfused kidneys of mice (Wagner et al. 2007). Moreover, intrarenal infusion of peptides known to block connexin40 (Cx40) or Cx37, both of which are expressed in JG cells (Arensbak et al. 2001; Haefliger et al. 2001; Hwan Seul and Beyer 2000; Zhang and Hill 2005), elevated plasma renin activities in rats in vivo (Takenaka et al. 2008). Supportive data for a role of connexins in the control of renin release come from studies in genetically modified mice. Mice in which the coding region for Cx43 has been replaced by Cx32 (Cx32KI mice) have lower PRC and renal renin gene expression levels than wild-type mice (Haefliger et al. 2006). Moreover, neither the inhibitory effect of a high-salt diet on renin gene expression nor the increase in PRC and renin mRNA levels induced by renal hypoperfusion were observed in Cx32KI mice (Haefliger et al. 2006). Although these data clearly suggest a functional role of connexins in the regulation of renin release, how Cx43 or Cx32 interferes with the renin system remains unknown, since neither isoform has yet been shown to be expressed in JG cells.

Moreover, mice with a genetic deletion of Cx40, which is the major connexin in JG cells (Arensbak et al. 2001; Haefliger et al. 2001; Hwan Seul and Beyer 2000; Zhang and Hill 2005), have elevated plasma renin concentrations and are severely hypertensive (de Wit et al. 2003; Krattinger et al. 2007; Wagner et al. 2007). The inhibitory effect of high renal perfusion pressure or Ang II on renin release and PRC that occurs in wild-type mice is markedly attenuated in Cx40 KO mice in vivo and in vitro (de Wit et al. 2003; Krattinger et al. 2007; Wagner et al. 2007). Therefore, these data are consistent with the pharmacological studies mentioned above, indicating a functional role for gap junctions and specifically Cx40 in the control of renin release. The results with Cx40 KO mice should be interpreted with caution, however, since the typical expression pattern and morphology of JG cells is significantly altered in Cx40 KO mice. Renin-expressing cells in

Cx40 KO mice have a mesenchyme-like irregular shape and are not located in the vessel wall of afferent arterioles, but instead in the extraglomerular mesangium and in the periglomerular and peritubular interstitium (Kurtz et al. 2007). Therefore, the aberrant expression of JG cells in Cx40 KO mice may contribute to the disturbances in regulation of renin release in these mice.

The profound disturbance in the regulation of the renin system in Cx40 KO mice, as well as the pharmacological studies on the role of gap junctions, calls for efforts to identify the signal that is transferred by holo- or hemichannels formed by connexins. A hint may come from the fact that the inhibitory effect of blood pressure and Ang II on renin release depends on Ca^{2+} (Scholz et al. 1994; Schweda and Kurtz 2004). In fact, the regulation of renin release by perfusion pressure in isolated kidneys perfused with a nominally Ca^{2+}-free perfusate resembles Cx40-deficient kidneys or wild-type kidneys treated with a gap junction blocker (Wagner et al. 2007). Moreover, the reduction of extracellular Ca^{2+} concentration stimulates renin release from wild-type kidneys, but not from Cx40 KO kidneys (Wagner et al. 2007). These data therefore imply that Cx40 channels may interfere with Ca^{2+}-dependent processes or may be involved in Ca^{2+} handling of JG cells. Consistent with this hypothesis, the spreading of Ca^{2+} waves has been observed in the JG apparatus and in the renin-producing cell line As4.1. In microperfused JG apparatuses, increases in salt load at the macula densa induce a Ca^{2+} wave and elevate the intracellular Ca^{2+} concentration in the renin-producing JG cells (Peti-Peterdi 2006). The spreading of this Ca^{2+} wave is interrupted either by inhibition of extracellular ATP signaling or by blockade of gap junctional coupling (Peti-Peterdi 2006), which is consistent with previous reports that connexins enhance release of cellular ATP (Arcuino et al. 2002; Cotrina et al. 1998) or release ATP directly by functioning as hemichannels (Evans et al. 2006). Furthermore, in the renin-producing cell line As4.1, an ATP-dependent Ca^{2+} wave was observed in response to mechanical stress, thereby synchronizing the intracellular Ca^{2+} response in the cells (Yao et al. 2003). Since intracellular Ca^{2+} suppresses renin release, it is conceivable that a disturbance of the propagation and synchronization of Ca^{2+} signals in JG cells, which may result from Cx40 deficiency or blockade of gap junctions, accounts for the defective regulation of renin release in Cx40 KO mice.

Locally acting factors that regulate renin release

Renin release is under the complex control of numerous local and systemic factors (Table 1). For instance, the neuropeptides vasoactive intestinal polypeptide (VIP) and calcitonin gene-related peptide (CGRP) stimulate renin secretion (Gnaedinger et al. 1989; Kurtz et al. 1988c; Porter et al. 1982, 1983). These neuropeptides are expressed in nerve fibers surrounding the renal cortical vasculature in the vicinity of the renin-producing JG cells (Barajas et al. 1983; Kurtz et al. 1988c; Reinecke and Forssmann 1988), and they may therefore participate in the physiological control of renin release. Recently, the neuropeptide PACAP, which is highly homologous to VIP, has been shown to stimulate renin release from isolated perfused kidneys, cultured JG cells, and single JG cells (Hautmann et al. 2007). Similar to its effect

Table 1 Hormones regulating renin release

Stimulating hormones	
ß-adrenergic agonists	(Gnaedinger et al. 1989; Kurtz et al. 1988c)
Adrenomedullin	(Gnaedinger et al. 1989; Kurtz et al. 1988c)
Calcitonin gene-related peptide CGRP	(Gnaedinger et al. 1989; Kurtz et al. 1988c)
Dopamine	(Imbs et al. 1975; Kurtz et al. 1988b)
Glucagon	(Keeton and Campbell 1980; Vikse et al. 1985)
Parathyroid hormone	(Keeton and Campbell 1980; Saussine et al. 1993)
Pituitary adenylate cyclase-activating polypeptide PACAP	(Hautmann et al. 2007)
Prostaglandin E_2	(Jensen et al. 1996; Keeton and Campbell 1980)
Prostaglandin I_2	(Jensen et al. 1996; Keeton and Campbell 1980)
Vasoactive intestinal peptide VIP	(Porter et al. 1982; Porter et al. 1983)
Inhibiting hormones	
Adenosine	(Churchill and Churchill 1985; Weihprecht et al. 1990)
α-adrenergic agonists	(Keeton and Campbell 1980)
Angiotensin II	(Michelakis 1971; Vander and Geelhoed 1965)
Arginine-vasopressin	(Bunag et al. 1967; Kurtz et al. 1986c)
Atrial natriuretic peptide (ANP)	see chapter 6.1.
Endothelins	(Kurtz et al. 1991; Rakugi et al. 1988)
Neuropeptide Y (NPY)	(Bischoff et al. 1997; Hackenthal et al. 1987)

in other cell types, PACAP markedly increases cAMP levels in cultured renin-producing JG cells (unpublished observation). Consistent with the stimulatory effect of PACAP in vitro, mice lacking the PACAP receptor PAC1 have lower PRCs than their wild-type counterparts. Although PRC can be stimulated in PAC1 KO mice with a low-salt diet or ACE inhibitors, PRC levels are lower in KO mice than in their wild-type counterparts for each of the treatments (Hautmann et al. 2007), indicating that PACAP acts as a tonic enhancer of renin release. Thus, it appears likely that VIP and CGRP act as modulators of renin release in a similar way as PACAP. Further research is needed to test this hypothesis in vivo.

Atrial natriuretic peptide

ANP is a peptide hormone that is released primarily from the cardiac atria in response to an increase in wall tension due to elevated atrial pressure. ANP has numerous, well-established cardiovascular and renal effects that lead to natriuresis and decrease blood pressure (reviewed in Brenner et al. 1990; Kuhn 2005; Melo et al. 2000; Silver 2006). Thus, ANP counteracts the renin-angiotensin-aldosterone system that stabilizes blood pressure and stimulates renal salt reabsorption. In addition to this functional antagonism, ANP inhibits renal renin release. In the vast majority of studies, the application of exogenous ANP at concentrations near or above the upper limit of physiological levels reduced the plasma renin activity (for instance, see Anderson et al. 1987; Brands and Freeman 1988, 1989; Brown and O'Flynn 1989; Burnett et al. 1984; Carson et al. 1990; Cuneo et al. 1987; Garcia et al. 1985; Maack et al. 1984; Obana et al. 1985; Richards et al. 1988; Salazar et al. 1986; Struthers et al. 1986; Tomura et al. 1990). In other reports, however, ANP did not suppress plasma renin levels or renin release (Ehmke et al. 1992; Freestone et al. 1989; Grandis et al. 1992; Hirata et al. 1987; Ishimitsu et al. 1992; Pham et al. 1997).

Despite these inconsistencies, which may be partially explained by different protocols, methods of application, and doses of ANP (Bie et al. 1988; Brands and Freeman 1989; Ehmke et al. 1992), these studies indicate that ANP suppresses renin release in vivo under most circumstances. Although not unequivocally shown (Hiruma et al. 1986; Ishimitsu et al. 1992; Itoh et al. 1987; Rodriguez-Puyol et al. 1986; Takagi et al. 1988a), the ability of ANP to inhibit renin secretion has been corroborated in numerous in vitro studies using different models, including isolated JG cells (Henrich et al. 1987, 1988, 1986; Kageyama and Brown 1990; Kurtz 1986; Kurtz et al. 1986a, 1986b; Narumi et al. 1987). In general, ANP exerts its biological effects by activating guanylate cyclase A (GC-A), thereby inducing cGMP formation. In JG cells ANP also increases cGMP levels, which in turn suppresses renin release (Kurtz et al. 1986b).

In addition to these direct effects of ANP on JG cells inhibiting renin release, ANP can indirectly affect renin release through its effects on renal tubular transport processes. Thus, inhibition of tubular salt reabsorption by ANP can increase the sodium chloride load at the distal tubule including the macula densa. As discussed below, an increase in the NaCl concentration at the macula densa suppresses renin release from the JG cells in the same nephron. These indirect effects of ANP may contribute significantly to the suppression of renin release, given the results of studies on nonfiltering kidneys in dogs, some of which showed that ANP did not suppress renin release in the absence of tubular salt load (Deray et al. 1987; Opgenorth et al. 1986; Villarreal et al. 1986).

As mentioned above, ANP lowers blood pressure. Since blood pressure and renin release are inversely related (see below), ANP may also influence renin secretion through its hemodynamic effects. The complex interplay among ANP, blood pressure, and renin release may help to explain recent results obtained in ANP KO mice and GC-A KO mice. Based on the inhibitory effects of ANP on the renin system, these KO strains should show disinhibition of renin synthesis and release. However, ANP KOs and GC-A KOs have lower renal renin gene expression and Ang II levels than their respective wild types; plasma renin levels are reduced in GC-A KOs but are increased in ANP KOs compared to their wild-type counterparts (Holtwick et al. 2002; Melo et al. 1998; O'Tierney et al. 2008; Shi et al. 2001). Although the differences in plasma renin levels between ANP and GC-A KOs cannot be explained with available data, the results overall imply that ANP stimulates rather than inhibits renin synthesis and release. However, because of the systemic effects of ANP, both KO strains are severely hypertensive. Since a high renal perfusion pressure inhibits the renin system, it may be that the high blood pressure indirectly mediates the inhibition of the renin system. Consistent with this notion, newborn GC-A KO mice show increased renin gene expression levels, which are suppressed as blood pressure increases during the first weeks of life (Shi et al. 2001).

Taken together, the available data indicate that ANP suppresses renin release directly at the JG cell level, as well as indirectly through its tubular effects. These inhibitor effects can be modulated by the profound cardiovascular effects of ANP in vivo.

Angiotensin II

Ang II suppresses renin release, thereby establishing a negative feedback loop. Numerous in vivo and in vitro studies have unequivocally shown strong inhibition of renin release by Ang II. This effect can occur independently of tubular function or increases in renal perfusion pressure, suggesting the existence of rather direct effects on the renin-producing JG cells (Hackenthal et al. 1990; Kurtz et al. 1986c; Kurtz and Wagner 1999; Vander and Geelhoed 1965). This view is strengthened by the fact that JG cells are richly equipped with angiotensin AT_1 receptors and that Ang II suppresses renin synthesis and release in isolated JG cells. As discussed in the section entitled "Calcium," Ang II increases the cytosolic Ca^{2+} concentration in JG cells and inhibits renin release in a Ca^{2+}-dependent manner. The inhibitory effects of Ang II on renin synthesis and renin release are not only observed in studies in which Ang II is administered exogenously, they become even more evident when endogenous Ang II generation or Ang II signaling is interrupted. Thus, ACE inhibitors or AT_1 antagonists strongly increase PRC and expression of renal renin mRNA and protein, and induce recruitment of renin-expressing cells in the afferent arterioles.

In addition to its direct cellular effects, Ang II indirectly inhibits renin release. For instance, Ang II is a potent vasoconstrictor, and therefore it raises blood pressure. Since an increase in blood pressure per se suppresses renin release, the hemodynamic effects of Ang II further contribute to the suppression of renin release. According to two recent studies using chimeric mice carrying regional deletions of the AT_{1a} receptor, as well as cross-transplant studies using AT_{1a} receptor KO mice, interruption of Ang II signaling strongly stimulates renin synthesis and release rather indirectly: The stimulation is due to the reduction in blood pressure rather than to a disinhibition directly at the JG cells (Crowley et al. 2005; Matsusaka et al. 1996).

These studies make clear that Ang II acts through both direct and indirect mechanisms to potently inhibit renin release.

Prostaglandins E_2 and I_2

The formation of prostaglandins is initiated by the conversion of arachidonic acid to the intermediate prostaglandin H_2 (PGH_2). This reaction is catalyzed by cyclooxygenases, which exist in two isoforms, COX-1 and COX-2, both of which are expressed in the kidney (Hao and Breyer 2007; Harris et al. 1994). PGH_2 can be converted to several different prostaglandins, such as prostaglandin E_2 (PGE_2) and I_2 (PGI_2, prostacyclin), by corresponding prostaglandin synthases.

PGE_2 and PGI_2 stimulate renin release in a variety of models, including cultures of JG cells and single JG cells (Franco-Saenz et al. 1980; Friis et al. 2005; Gerber et al. 1979b; Hackenthal et al. 1980; Henrich and Campbell 1984; Ito et al. 1989; Jensen et al. 1996; Schweda et al. 2004b). Conversely, it has been known for a long time that blockade of prostaglandin synthesis can reduce plasma renin activity (PRA) in vivo

(Bolger et al. 1976; Vander 1968). In general, PGI_2 exerts its effects via one receptor type (IP receptor), while PGE_2 can exert its biological actions via four G protein-coupled receptor subtypes, named prostaglandin E1–4 receptors (EP1–4) (Breyer and Breyer 2000). Recently, experiments in isolated perfused kidneys and pharmacological studies on isolated and single JG cells showed that both EP2 and EP4 receptors mediate the stimulation of renin release by PGE_2 (Friis et al. 2005; Schweda et al. 2004b). By stimulating AC activity through IP receptors, as well as through EP2 and EP4 receptors, PGI_2 and PGE_2 significantly elevate cAMP formation in JG cells, and stimulate renin exocytosis in a PKA-dependent manner (Friis et al. 2005; Jensen et al. 1996).

Interest in the role of PGE_2 and PGI_2 in the control of renin release picked up several years ago with the demonstration that COX-2 localizes within the macula densa and its expression changes in parallel with renin synthesis in JG cells in response to a variety of challenges (Cheng et al. 1999; Harris et al. 1994; Mann et al. 2001a, 2001b). In fact, as discussed below, COX-2 products exert a tonic enhancing effect on renin synthesis and secretion in vivo, underlining the stimulatory roles of PGE_2 and PGI_2 on renin release.

Nitric oxide

NO is synthesized from L-arginine by nitric oxide synthases (NOS), which exist in three isoforms: neuronal NOS (nNOS or NOSI), inducible NOS (iNOS or NOSII), and endothelial NOS (eNOS or NOSIII). Renin-producing JG cells are surrounded by cells that have the capacity to form NO. In addition to the endothelial cells of the afferent arteriole, which express eNOS, macula densa cells express nNOS (Bachmann et al. 1995; Mundel et al. 1992; Ujiie et al. 1994; Wilcox et al. 1992). Similar to what has been shown for COX-2, nNOS localizes within the macula densa and its expression changes in parallel with renin expression in response to renal hypoperfusion, salt intake, and loop diuretics (Bosse et al. 1995; Schricker et al. 1996; Singh et al. 1996; Tojo et al. 1995).

Because of the close proximity of renin-producing and NO-producing cells, several studies have examined the functional role of NO in the regulation of renin release. Despite some conflicting results, there is overall agreement that NO stimulates renin release (for detailed discussion, see Kurtz and Wagner 1998). As discussed above, NO likely enhances renin secretion by inhibiting PDE-3 activity, thereby blocking cAMP degradation and increasing intracellular cAMP levels. In order to define the relative importance of eNOS and nNOS in renin expression and release, researchers have investigated eNOS and nNOS KO mice: It has been found that eNOS KO mice show reduced renal renin gene expression and renin content, although their PRC is normal or even slightly higher than that of their wild-type counterparts (Beierwaltes et al. 2002; Castrop et al. 2004b; Shesely et al. 1996; Wagner et al. 2000). These results might suggest that NO derived from eNOS stimulates renin synthesis, but not renin release in vivo. In contrast, nNOS KO mice have lower PRC and renal renin content than their wild-type counterparts (Castrop et al. 2004b; Harding et al. 1997), although their renin mRNA expression is not different (Wagner et al. 2000). This suggests that NO derived from nNOS primarily stimulates translation and release of renin.

In conclusion, NO stimulates renin release and acts as a tonic enhancer of the renin system. This view is corroborated by studies on salt-dependent regulation of renin release that are discussed below in this review (Beierwaltes 1997; Castrop et al. 2004b; Sällström et al. 2008; Tan et al. 1999).

The macula densa mechanism

In the juxtaglomerular apparatus, the distal tubule comes into close contact with the vascular pole of the parent glomerulus. Because of the morphological appearance of the cells in this tubular segment, this contact point has been termed "macula densa." The close spatial relationship of the macula densa to the afferent arteriole, which contains vascular smooth muscle cells and renin-producing JG cells, led to the assumption of a functional interrelationship between tubular fluid on the one hand and vascular tone of the afferent arteriole and renin release on the other hand. In fact, an increase in afferent vascular tone, and therefore a reduction in glomerular filtration due to a high NaCl concentration at the macula densa, was demonstrated convincingly, and led to the concept of the tubuloglomerular feedback mechanism (Castrop 2007; Schnermann and Levine 2003; Thurau and Schnermann 1965). Moreover, a functional role of the macula densa in the regulation of renin release has been demonstrated in elegant studies using a preparation of isolated, perfused JG apparatus. This work has shown that renin release increases in response to acute reductions in NaCl concentration at the macula densa (Skott and Briggs 1987). Therefore, an inverse relationship is assumed to exist between salt concentration at the macula densa and glomerular filtration and renin release.

The inhibition of renin release by high tubular NaCl concentration depends on the tubular Cl^- concentration, not on the Na^+ concentration (Lorenz et al. 1991). There is consensus that macula densa cells sense the luminal Cl^- concentration via the transport rate of the apical $Na^+/K^+/2Cl^-$ cotransporter (NKCC2), which is expressed in the tubular cells of the thick ascending limb of Henle (TALH) as well as in the macula densa cells. Pharmacological blockade of NKCC2 with loop diuretics interrupts the salt-dependent regulation of renin release in isolated, perfused JG apparatuses and afferent arterioles with attached macula densa (He et al. 1995; Itoh and Carretero 1985; Lorenz et al. 1991). A functional role of the macula densa in the control of renin release in vivo is suggested by studies showing that acute infusion of NaCl increases the Cl^- concentration in the early distal tubular fluid (Lorenz et al. 1990) and reduces renin release and PRA (Gerber et al. 1979a; Kim et al. 2006; Kirchner et al. 1978; Lorenz et al. 1990). Conversely, pharmacological blockade of NKCC2 with loop diuretics potently stimulates renin release in vivo. However, an unequivocal conclusion that the stimulation of renin release in response to loop diuretics results from the inhibition of macula densa salt transport is complicated primarily by two circumstances:

1. Loop diuretics block NKCC2 and NKCC1, a further isoform of the NKCC transporters (Hannaert et al. 2002). Since NKCC1 is expressed in JG cells (Kaplan et al. 1996), loop diuretics may act directly on JG cells. In fact, loop diuretics directly stimulate renin release from isolated JG cells in a NKCC1-dependent fashion (Castrop et al. 2005). Moreover, NKCC1 KO mice show elevated baseline PRC levels, suggesting an inhibitory role for NKCC1 on renin release in vivo (Castrop et al. 2005). Despite the functional role of NKCC1 in renin release, furosemide markedly raises PRC in both NKCC1 KO and wild-type mice in vivo. Therefore, the marked stimulation of renin release in response to loop diuretics in vivo is related at least predominantly to an inhibition of NKCC2.
2. Loop diuretics block NKCC2 activity not only in macula densa cells but also in the more upstream parts of the loop of Henle, resulting in marked salt and water loss. As a consequence, salt content of the body and the blood pressure decreases, and this induces systemic counter balancing effects, such as activation of the sympathetic nervous system. All of these systemic effects enhance renin release, as discussed in the next section. Therefore, it appears plausible that the stimulation of PRC by loop diuretics in vivo may be the indirect result of the diuretic effects rather than the direct result of the blockade of macula densa salt transport.

NKCC2 exists as three splice variants in the TALH, named NKCC2A, NKCC2B, and NKCC2F. NKCC2F is the dominant form, and it reabsorbs the greatest amount of NaCl in the medullary part of the TALH, while NKCC2A and NKCC2B are expressed in the downstream parts of the TALH, including the macula densa. NKCC2A and NKKC2B KO mice do not display the signs of salt deficiency that have been reported for classical NKCC2 KO mice that lack all three splice variants of NKCC2 (Oppermann et al. 2006, 2007; Takahashi et al. 2000). This finding underscores the critical role of the NKCC2F splice variant for salt reabsorption in the TALH. Moreover, it indicates that NKCC2A and NKCC2B KO mice are useful tools for investigating the role of NKCC2 salt transport in the macula densa in regulating renin release without the confounding variables of salt and water loss. Under control conditions, PRC levels are similar between NKCC2A or NKCC2B mice and their wild-type counterparts (Oppermann et al. 2006, 2007). However, the reduction of PRC in response to an acute intravenous salt load was markedly attenuated in NKCC2A KO mice (Oppermann et al. 2007), indicating that the salt transport rate of NKCC2A is important in the series of signaling events connecting tubular salt load to the release of renin from JG cells.

In addition to the NKCC2 transporter, the Na^+/H^+ exchanger NHE2 contributes significantly to salt transport in macula densa cells (Peti-Peterdi et al. 2000). NHE2 is located in the apical membrane of macula densa cells, and it links cell volume and intracellular pH to the Na^+ concentration in the distal tubular fluid. In line with a functional role for NHE2 in the macula densa control of renin release, NHE2 KO mice show increased renal renin content and activity and elevated PRC (Hanner et al. 2008). Moreover, a low-salt diet for 1 week increases renin content and activity in wild-type mice, but not in NHE2 KO mice, suggesting a functional role for this exchanger in the salt-dependent regulation of renin synthesis (Hanner et al. 2008). Given the substantial effects of NHE2 in the salt-dependent regulation of the renin system, which appear to be mediated via an upregulation of the prostaglandin forming Cox-2, further studies are needed to explore the role of this exchanger in other control mechanisms of renin release, such as its pressure dependence.

While the salt transport rate at the macula densa cells is generally accepted to be the initial sensing step in the macula densa control of GFR and renin release, the further downstream signaling connecting macula densa cells to their respective effector cells has been less clear. In the tubuloglomerular feedback mechanism, ATP and adenosine form a cascade of signaling molecules that adjust vascular tone to the salt concentration at the macula densa (Castrop 2007; Schnermann and Levine 2003). Macula densa cells release ATP in response to an increase in extracellular NaCl concentration (Bell et al. 2003; Komlosi et al. 2004), and ATP can be degraded in the interstitium of the JG apparatus into adenosine, which constricts the afferent arteriole by activating A1 adenosine receptors (A1AR) (Fig. 3). The functional importance of this sequence of events has been elaborated in several studies using pharmacological or KO strategies. Thus, interruption of adenosine signaling via blockade or deletion of A1AR, as well as of adenosine formation via blockade or deletion of ecto-5'-nucleotidase, prevented or at least significantly impaired the tubuloglomerular feedback

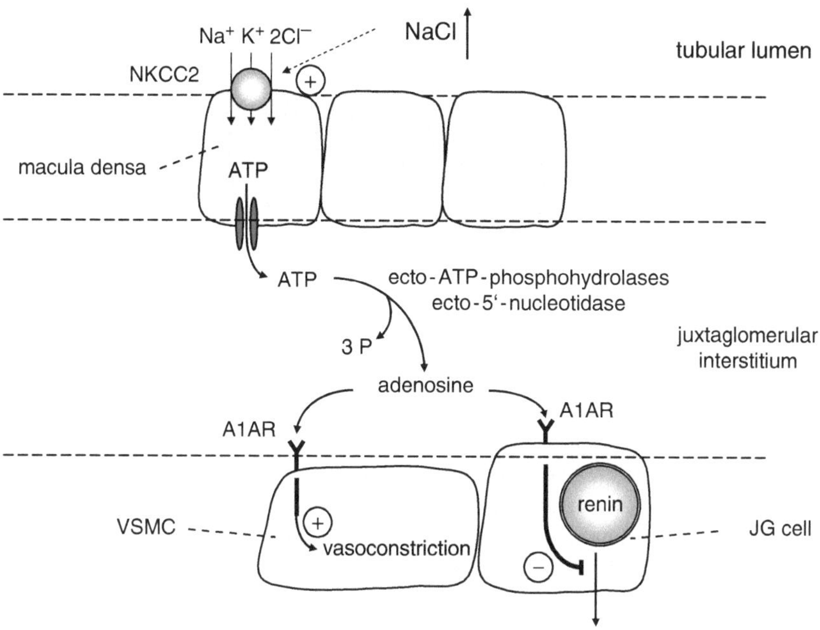

Fig. 3 Regulation of renin release and vascular tone of the afferent arteriole in response to an increase in the tubular NaCl concentration. Macula densa cells sense tubular chloride concentration ([NaCl]) by the transport activity of the NKCC2 transporter. ATP is released from macula densa cells in response to an increase in [NaCl]. ATP can be degraded in the juxtaglomerular interstitium to adenosine. Adenosine inhibits renin release and induces contraction of vascular smooth muscle cells (*VSMC*) via activation of A1 adenosine receptors (*A1AR*)

response (Brown et al. 2001; Castrop et al. 2004a; Ren et al. 2002; Schnermann et al. 1990; Sun et al. 2001; Thomson et al. 2000; Wilcox et al. 1999). Based on the fact that adenosine inhibits renin release through A1AR (Albinus et al. 1998; Kurtz et al. 1988a; Weihprecht et al. 1990), researchers speculated that the same signaling events mediate the macula densa control of renin release. In fact, blockade of A1AR significantly attenuated the inhibition of renin release from isolated, perfused JG apparatus preparations in response to a high NaCl concentration at the macula densa (Weihprecht et al. 1990). Moreover, the injection of isotonic saline significantly suppressed PRC in wild-type mice, but not in mice lacking the A1 adenosine receptors (Kim et al. 2006). Notably, the stimulation of PRC and renin release by loop diuretics, which blocks macula densa salt transport as discussed above, was similar between A1AR KO mice and wild-type mice in vivo and in isolated perfused kidneys (Kim et al. 2006; Schweda et al. 2003). Therefore, these data suggest that adenosine and its receptor A1AR are critical for inhibiting renin release in response to an acute increase in tubular NaCl concentration, but they are not critical for stimulating renin release induced by a blockade of macula densa salt transport.

Taken together, there is substantial evidence, derived mainly from in vitro and in vivo studies using acute experimental interventions, to suggest that the macula densa regulates renin release in response to acute changes in the tubular NaCl concentration. This has also led researchers to hypothesize that the macula densa mechanism regulates renin synthesis and renin release by the salt intake in the long term, too. However, convincing experimental evidence for this idea remains elusive, as discussed below.

Renal baroreceptor mechanism

Renal renin release is inversely related to renal perfusion pressure such that a drop in perfusion pressure stimulates renin secretion, whereas an increase in perfusion pressure inhibits it. This so-called renal baroreceptor mechanism has been documented in several species and experimental models. The precise signal-effector coupling between renal arterial pressure and renin secretion has not yet been characterized precisely. In contrast to the early assumption that the pressure-dependent regulation of renin release occurs via the macula densa mechanism, it is now widely accepted that blood pressure changes are detected within the afferent arteriole itself and pressure-dependent regulation of renin release is not absolutely dependent on macula densa function.

This conclusion is based on several lines of evidence. First, regulation of renin release by renal perfusion pressure is preserved in nonfiltering kidneys in vivo and in vitro (Blaine et al. 1971, 1970; Scholz et al. 1993). Second, stepwise reduction of renal perfusion pressure in dogs stimulates renin release before renal blood flow or glomerular filtration are reduced (Kirchheim et al. 1987). Third, the pressure dependent regulation of renin release is preserved even in isolated, perfused glomeruli lacking a functional macula densa (Bock et al. 1992). Finally, pressure-dependent regulation of renin release is observed in isolated JG cells and renin expressing cell lines. As a tool to mimic an increase in blood pressure, cells were stretched (Carey et al. 1997; Ryan et al. 2000) or a high transmural pressure was applied (Hirota et al. 2002; Ichihara et al. 1999). In fact, both maneuvers significantly inhibit renin release from JG cells (Carey et al. 1997; Hirota et al. 2002; Ichihara et al. 1999). Therefore, assuming that the stimulation protocols mimic changes in perfusion pressure, these data appear to indicate that the baroreceptor resides in the renin-producing cells themselves. Notably, mechanical distension induces a Ca^{2+} influx and an activation of phospholipase C (PLC), resulting in elevated inositol phosphate levels in the renin-producing cell line As4.1 (Ryan et al. 2000). In parallel, the inhibition of renin release due to transmural pressure is prevented in the absence of extracellular Ca^{2+} or pharmacological blockade of PLC (Hirota et al. 2002; Ichihara et al. 1999). Such Ca^{2+} dependence in the pressure dependent regulation of renin release has also been observed in isolated perfused kidneys of rats and mice (Scholz et al. 1994; Wagner et al. 2007). This raises the possibility that the Ca^{2+} influx pathway is involved in this process. On the one hand, Ca^{2+} influx may be directly induced by the activation of stretch-induced Ca^{2+} channels. On the other hand, JG cells may release signaling molecules in response to an increase in pressure and these molecules may induce Ca^{2+} influx and Ca^{2+}-dependent inhibition of renin release in an autocrine or paracrine fashion. Among others, ATP and adenosine are possible candidates for the latter pathway. In JG cell cultures, mechanical stimulation of one cell induces the spreading of a Ca^{2+} wave that is dependent on extracellular ATP (Yao et al. 2003). Moreover, JG cells release ATP when they are mechanically stimulated (Yao et al. 2003). Alternatively, as discussed above in detail, the interstitium of the JG apparatus has the capacity to degrade ATP into adenosine. Therefore, ATP release may result in an increase in adenosine, ultimately inhibiting renin release via the activation of A1 adenosine receptors (A1AR). Some support for this hypothesis comes from studies demonstrating that the pressure-dependent inhibition of renin release is completely absent in isolated perfused kidneys from A1AR KO mice (Schweda et al. 2005). Moreover, the suppression of PRC by a pharmacologically induced blood pressure increase is markedly attenuated in A1AR KO mice and mice lacking 5'-ectonucleotidase, an enzyme responsible for conversion of AMP into adenosine (Oppermann et al. 2008; Schweda et al. 2005). Notably, in studies with A1AR KO mice, a drop in perfusion pressure in isolated perfused kidneys or a reduction in blood pressure by hydralazine in vivo stimulated renin release as in the wild-type controls, indicating that adenosine and A1AR are involved only in the inhibition of renin release by an increase but not in the stimulation of renin release by a decrease in perfusion pressure (Kim et al. 2006; Schweda et al. 2005).

Taken together, there are several lines of evidence for a renal baroreceptor mechanism, which may reside in the JG cells themselves. The molecular nature of this baroreceptor for detecting changes of perfusion pressure and inducing downstream signaling has yet to be determined precisely. Moreover, as discussed in the following section, the regulation of renin release by blood pressure at the systemic level is significantly modulated by factors additional to the renal baroreceptor mechanism.

Systemic factors controlling renin release

In addition to locally acting factors effecting renin release from the JG cells, renin release is also controlled systemically by salt intake and renal perfusion pressure in negative feedback loops (Hackenthal et al. 1990). Moreover, the sympathetic nervous system is a central regulator of renin release from JG cells, as has been unequivocally demonstrated in numerous studies using pharmacological approaches, renal nerve stimulation, or renal denervation (for review see DiBona and Kopp 1997; Hackenthal et al. 1990; Kurtz 1989; Wagner and Kurtz 1998). Further support for a critical role of catecholamines and β-adrenergic receptors in the control of renin release comes from a recent study in β_1/β_2-adrenoreceptor double KO mice (Kim et al. 2007a). This model has an advantage over pharmacological approaches, since β-adrenoreceptors are completely absent; it also has an advantage over renal denervation approaches, since it is free of the effects not only of renal nerves but also of circulating catecholamines. The PRC of β_1/β_2 KO mice is strongly reduced to 15% of the level of the wild-type controls. In addition, the increase in PRC in response to different stimuli is markedly attenuated (Kim et al. 2007a), underscoring the importance of sympathetic nervous tone as a major control mechanism of renin release.

Blood pressure

Because of the activation of the renin-angiotensin-aldosterone cascade, stimulation of renin release is accompanied by an increase in blood pressure. Since blood pressure influences renin secretion in an inverse fashion, a negative feedback loop is established between blood pressure and renin release. As mentioned above, there is convincing evidence that pressure changes can be detected by the afferent arteriole and likely by the JG cells themselves. Nevertheless, these direct effects can be significantly modulated by several intrarenal and systemic factors. For instance, the macula densa, which is not essential for the renal baroreceptor mechanism, can contribute to the pressure-dependent regulation of renin release (Scholz et al. 1993; Schricker et al. 1994c). Moreover, renal autacoids, such as NO and prostaglandins, significantly modulate pressure-dependent regulation of renin release. Nonselective blockade of all NO synthases markedly attenuates the stimulation of renin secretion or PRA by a low renal perfusion pressure in isolated perfused kidneys and in rats with renal artery stenosis (Knoblich et al. 1996; Scholz and Kurtz 1993; Schricker et al. 1994a, 1994b). These results indicate a tonic stimulatory role of NO in the pressure-dependent regulation of renin release (Scholz and Kurtz 1993). Since the stimulation of PRC by a reduction in renal perfusion pressure is not attenuated in eNOS KO mice, it can be speculated that nNOS is the major source of NO in this context (Beierwaltes et al. 2002).

Prostaglandins have also been implicated in the regulation of renin release by blood pressure. Both PGE_2 and PGI_2 are potent, direct stimulators of renin release

(Jensen et al. 1996), and they are released from the kidneys in response to a reduced perfusion pressure in dogs and humans (Imanishi et al. 2001; Jackson et al. 1982; Vikse and Kiil 1985). In fact, studies using nonselective blockade of cyclooxygenase activity indicated a modulating role of cyclooxygenase products in the blood pressure-dependent regulation of the renin system (Hackenthal et al. 1990). The interest in the role of prostanoids was reinforced by the correlation between renal perfusion pressure and COX-2 expression in the macula densa (Hartner et al. 1998; Mann et al. 2001a). Studies on the functional role of prostaglandins derived from COX-2 have given conflicting results: While some studies found COX-2 to be involved in the stimulation of renin release and synthesis in response to renal hypoperfusion or a drop in arterial blood pressure (Fujino et al. 2004; Kim et al. 2007b; Wang et al. 1999), other studies did not (Hartner et al. 2003; Mann et al. 2001a; Richter et al. 2004).

These data indicate that prostanoids derived from COX-2 may have a modulatory effect on the pressure-dependent regulation of renin release. Notably, the increase in blood pressure, plasma renin activity (PRA), and renal renin expression in response to unilateral renal artery stenosis is significantly attenuated in mice lacking the PGI_2 receptor IP (Fujino et al. 2004), indicating that PGI_2 is the relevant prostanoid.

A recent study in rats using the selective COX-2 blocker SC-58236 may help to explain some of the conflicting findings mentioned above. Although a single dose of SC-58236 significantly blocked prostaglandin formation, it did not suppress PRA under baseline conditions or in response to unilateral renal artery stenosis (Matzdorf et al. 2007). However, after 5 days of treatment, once renin synthesis and PRA were already reduced at baseline, the stimulation of PRA in response to subsequent renal artery clipping was significantly attenuated (Matzdorf et al. 2007). Thus, these data indicate that prostaglandins derived from COX-2 act as enhancers, but not critical mediators, of renin release in response to renal hypoperfusion. This effect occurs via an activation of renin synthesis that is the basis for the overall stimulability of renin release.

Finally, renal sympathetic nerves have a significant impact on pressure-dependent renin release. At normal systemic blood pressures and blood volumes, afferent input from cardiopulmonary and carotid baroreceptors inhibits renal sympathetic nerve activity and therefore renin release (Thrasher 1994). Thus, a drop of arterial pressure unloads carotid baroreceptors and stimulates renin release (Thrasher 1994; DiBona and Kopp 1997). Besides these effects, renal sympathetic nerves augment the renal baroreceptor mechanism. Thus, studies using renal nerve stimulation, renal denervation, or pharmacological approaches have shown that sympathetic nervous tone enhances renin release in response to a low renal perfusion pressure (Fan et al. 1994a, 1994b; Wagner et al. 1999; see review in Hackenthal et al. 1990).

Taken together, blood pressure controls renin release through at least two mechanisms. On one hand, renal sympathetic nerve activity that significantly affects renin release is regulated by systemic blood pressure and blood volume through carotid and cardiopulmonary baroreceptor reflex loops. On the other hand, changes in systemic blood pressure result in parallel alterations in renal perfusion pressure that control renin release via the intrarenal baroreceptor mechanism. The gain in renin response upon renal baroreceptor activation is significantly modulated by the macula densa mechanism, NO, prostaglandins, and the sympathetic nervous system. In this way, blood pressure regulates renin release through a complex web of interacting conditions and molecules.

Regulation of renin release by salt load

It has been known for a long time that the NaCl balance of the body is a major determinant of renin synthesis and release: Low salt intake stimulates it, while high salt intake inhibits it. Following the elegant studies demonstrating that a reduction in tubular NaCl concentration at the macula densa stimulates renin release, researchers hypothesized that macula densa salt transport (see above) links the salt load and salt intake to renin synthesis and renin release. However, as discussed in this section, considerable evidence has cast doubt on the notion that the macula densa mechanism plays a central mediator role in regulating the renin system at least in response to changes in salt intake in the long term.

Macula densa cells express nNOS and COX-2, and the expression of both enzymes changes in parallel with salt intake and is stimulated by blockade of NKCC2 transport activity (Bachmann et al. 1995; Bosse et al. 1995; Harris et al. 1994; Mann et al. 2001b; Mundel et al. 1992; Schricker et al. 1996; Singh et al. 1996; Tojo et al. 1995; Wilcox et al. 1992). Moreover, macula densa cells release PGE_2 in response to low salt concentration (Peti-Peterdi et al. 2003; Yang et al. 2000b). Since the products of both enzymes, namely, NO and prostaglandins, stimulate renin release (see the sections entitled "Prostaglandins E_2 and I_2" and "Nitric Oxide"), it was speculated that nNOS or COX-2 products mediate salt-dependent regulation of renin release.

In fact, nonselective blockade of NO synthesis attenuates the stimulation of renin mRNA and PRA in response to a low-salt diet or to loop diuretics (Castrop et al. 2004b; Reid and Chou 1995; Schricker and Kurtz 1996; Tharaux et al. 1997). In order to determine the relative contribution of eNOS, which is expressed in endothelial cells of the afferent arteriole in close vicinity to the JG cells (Bachmann et al. 1995; Ujiie et al. 1994), studies have investigated the regulation of the renin system in eNOS KO mice. In these mice, loop diuretics, as well as different salt diets, regulate PRC and renin synthesis in the same way as in wild-type mice, indicating that eNOS is not essential for the salt-dependent regulation of the renin system (Castrop et al. 2004b; Wagner et al. 2000). However, the data suggest that NO derived from eNOS exerts a tonic stimulatory effect on renin release, since the absolute levels of PRC and renin gene expression were lower in eNOS KO mice than in their wild-type counterparts under the different treatments (Castrop et al. 2004b; Wagner et al. 2000).

Several studies also indicate a functional role for nNOS. In some studies, blockade of nNOS activity with 7-nitroindazole (7-NI) in rats or genetic deletion of nNOS in mice attenuated or abrogated the stimulation of renin release, PRC, and renal renin synthesis induced by a low-salt diet (Beierwaltes 1997; Sällström et al. 2008; Tan et al. 1999). In another study, however, this blockade did not attenuate the stimulation of renin content (Harding et al. 1997). In contrast to the low-salt diet treatment, renal renin mRNA levels and renin content were similar between nNOS KO mice and their wild-type counterparts when salt intake was high (Sällström et al. 2008; Wagner et al. 2000). Moreover, the suppression of PRC by acute Na^+ loading was not attenuated by pharmacological blockade of nNOS (Kompanowska-Jezierska et al. 2008), suggesting that NO derived from nNOS has no functional role in the regulation of the renin system under high salt load. Finally, loop diuretics that block tubular and macula densa salt transport and induce renal salt loss stimulated PRC in nNOS KO mice, but not in anesthetized rats treated with 7-NI (Beierwaltes 1995; Castrop et al. 2004b).

Some of these conflicting results may be explained by the different experimental conditions and by the unclear uptake of 7-NI in macula densa cells. The selectivity of 7-NI for nNOS over eNOS results from a limited uptake of the drug into endothelial cells compared to neurons (Moore and Bland-Ward 1996). Since the uptake characteristics of 7-NI into macula densa cells are unknown, care should be taken in considering 7-NI as a selective blocker of macula densa nNOS. Despite these limitations and the inconsistencies mentioned above, it seems reasonable to conclude that NO derived from nNOS can significantly influence the stimulation of PRA in response to salt deprivation, at the very least by acting as a positive enhancer. Such modulatory roles for NO derived from nNOS and eNOS, as well as NO in general, have also been reported in studies in isolated perfused kidneys of rats and mice (Castrop et al. 2004b; Kurtz et al. 1998b; Kurtz and Schweda 2006). Therefore, NO has a general permissive effect here by augmenting renin release in response to different stimuli and most likely also in response to a low salt intake.

Studies of the role of COX-2 in the regulation of renin release by salt intake have produced conflicting results. On one hand, stimulation of PRA or renin gene expression in response to low-salt diet or loop diuretics was abolished or at least attenuated after several days of pharmacological blockade of COX-2 or the genetic deletion of COX-2 (Fujino et al. 2004; Harding et al. 1997; Kammerl et al. 2001a, 2001b; Stichtenoth et al. 2005; Yang et al. 2000a). On the other hand, other studies using selective COX-2 blockers found no significant role for COX-2 (Höcherl et al. 2002b; Rodriguez et al. 2000). These confusing results may be partially explained by the results of two recent studies suggesting that prostanoids derived from COX-2 are involved in the control of renin synthesis and thereby set the baseline level and the gain in the general stimulation of renin release (Kim et al. 2007b; Matzdorf et al. 2007). Although a single dose of the selective COX-2 blocker SC-58236 reduces PGE_2 formation, it does not reduce PRA. However, treatment with SC-58236 for 5 days inhibits renin gene expression and PRA (Matzdorf et al. 2007). In this situation of low renin synthesis, the stimulation of PRA in response to a low-salt diet but also to various other stimuli, such as renal artery clipping, ACE inhibition, and the β-adrenoreceptor agonist isoproterenol, is markedly attenuated (Matzdorf et al. 2007). Therefore, these data suggest that prostaglandins derived from COX-2 control the overall stimulability of renin release but do not specifically interfere with the salt-dependent or macula densa control of renin release.

This same conclusion can be drawn from a study using mice deficient in COX-2 (Kim et al. 2007b): These mice have a lower PRC at baseline than their wild-type counterparts, and they show a smaller increase in the absolute value of PRC in response to furosemide, the vasodilator hydralazine, isoproterenol, blockade of ACE activity, or of angiotensin AT_1 receptors. However, the increases in PRC in response to these stimuli are similar with respect to the corresponding baseline levels in KO and wild-type mice, indicating a clear correlation between the PRC levels at baseline and after stimulation. Notably, when renin synthesis is prestimulated in COX-2-deficient mice over several days so that levels of renin mRNA and PRC reach those of wild-type mice, the two genotypes respond similarly to subsequent acute stimulation of PRC by the loop diuretic furosemide (Kim et al. 2007b). Together, these data suggest that prostanoids derived from COX-2 are not essential as specific mediators for macula densa or salt-dependent control of renin release, but they are involved in setting the baseline level of renin synthesis. In this way, these prostanoids regulate the pool of releasable renin and the overall stimulation of renin release in response to a variety of factors.

Since COX-2 and nNOS in the macula densa have modulatory roles but are not required for the salt-dependent regulation of renin release, it appears plausible to question the role of the macula densa per se in regulation of the renin system by salt intake. As discussed in the section entitled "Macula densa mechanism," several in vitro studies have shown that changes in NaCl concentration in the distal tubular fluid result in the opposite changes in renin release from the same nephron. In vivo, an acute injection of isotonic saline increased the Cl^- concentration in the distal tubular fluid and at the same time reduced PRA in rats (Lorenz et al. 1990). A functional role of the macula densa mechanism in the inhibition of renin release by acute salt load was corroborated in NKCC2A KO mice, which lack one of the two splice variants of the NKCC2 cotransporter that are expressed in macula densa cells and that act as salt sensors in the tubular fluid. In these mice, the suppression of PRC in response to acute saline injection is significantly attenuated (Oppermann et al. 2007). In contrast, the long-term regulation of PRC by high- or low-salt diet for 1 week is not attenuated in NKCC2A KO mice (Oppermann et al. 2007).

These conflicting findings between short- and long-term salt loading conditions may be explained by a compensation of the missing NKCC2A by NKCC2B or by another transport mechanism when the salt load is constantly changed. Alternatively, these data are compatible with the assumption that macula densa salt transport per se is not essential for the chronic salt-dependent regulation of the renin system.

Such a dissociation of acute and chronic regulation of PRC in response to salt load as observed in NKCC2A KO mice has also been observed in mice with a genetic deletion of the A1 adenosine receptor (A1AR). Adenosine is a potent vasoconstrictor and inhibitor of renin release that functions by activating A1AR (Albinus et al. 1998; Kurtz et al. 1988a; Weihprecht et al. 1990). Both the macula densa control of glomerular filtration and the inhibition of PRC through an acute salt load are abolished in A1AR KO mice (Kim et al. 2006). In contrast, although a high salt intake increases the adenosine concentration in the renal cortex (Osswald et al. 1980;

Siragy and Linden 1996; Zou et al. 1999), both wild-type and A1AR KO mice show similar inhibition of renin synthesis and PRC in response to a high-salt diet over 1 week (Brown et al. 2006; Schweda et al. 2003). Again, these data are compatible with the notion that acute and long-term changes of salt load regulate renin release through different mechanisms, suggesting that the long-term control of renin release does not absolutely require the macula densa mechanism. Support for this assumption comes from studies in rats or mice with one hydronephrotic, nonfiltering kidney in which glomerular filtration and tubular salt load are abrogated and significant tubular damage occurs. Since these nonfiltering kidneys retain the ability to regulate renal cortical expression of renin mRNA and protein in response to changes in salt intake over 1 week (Barrett et al. 1990; Schweda et al. 2004a; Zhang and Morgan 1994), the data suggest that extratubular factors independent of macula densa salt transport are capable of maintaining the salt-dependent regulation of renin synthesis.

Taken together, the data indicate that the macula densa mechanism, which detects short-term changes in the tubular Cl^- concentration, is critical for inhibiting renin release due to acute increases in tubular salt load, whereas clear experimental evidence for a role of the macula densa as a mediator of the long-term regulation of the renin system through changes in salt intake is still lacking.

If not the macula densa, what factors could mediate the regulation of renin release by salt intake? Studies showing that renal nerve activity and renal norepinephrine release increase during salt depletion have suggested that the inverse relation between salt intake and renin release is mediated by the sympathetic nervous system (DiBona and Sawin 1985; Friberg et al. 1990; Oliver et al. 1980). However, there is good evidence from studies using renal denervation, pharmacological, or KO approaches that sympathetic tone enhances the stimulation of renin release in response to a low-salt diet but is not absolutely required for it (Brubacher and Vander 1968; Fernandez-Repollet et al. 1985; Höcherl et al. 2002a; Holmer et al. 1993; Kim et al. 2007a; Kompanowska-Jezierska et al. 2008; Pieruzzi et al. 2002; Tkacs et al. 1990). Similarly, the neuropeptide PACAP acts as a tonic enhancer of renin release during a low-salt diet (Hautmann et al. 2007).

Furthermore, since blood pressure can change in parallel with salt intake, it appears plausible that changes in renal perfusion pressure indirectly mediate the salt-dependent regulation of renin release. However, no studies supporting this conclusion have yet been published. Instead, several studies have reported salt-dependent regulation without accompanying changes in blood pressure (Brubacher and Vander 1968; Holmer et al. 1993).

Therefore, we must conclude that, although adjustment of renin release to salt load is a central regulating mechanism of the salt and water homeostasis of the body, the central switch that adjusts renin synthesis and release to the salt intake is not entirely clear. While acute increases in NaCl load inhibit renin release due to the involvement of the macula densa, the signaling pathways connecting long-term salt intake to renin synthesis and release remain to be determined. However, it should be emphasized that prostaglandins derived from COX-2, NO, and the sympathetic nervous system act as positive enhancers to significantly modulate salt-dependent renin release.

References

Abel KJ, Gross KW (1990) Physical characterization of genetic rearrangements at the mouse renin loci. Genetics 124:937-47

Albinus M, Finkbeiner E, Sosath B, Osswald H (1998) Isolated superfused juxtaglomerular cells from rat kidney: a model for study of renin secretion. Am J Physio l275:F991-7

Anderson JV, Donckier J, Payne NN, Beacham J, Slater JD, Bloom SR (1987) Atrial natriuretic peptide: evidence of action as a natriuretic hormone at physiological plasma concentrations in man. Clin Sci (Lond) 72:305-12

Arcuino G, Lin JH, Takano T, Liu C, Jiang L, Gao Q, Kang J, Nedergaard M (2002) Intercellular calcium signaling mediated by point-source burst release of ATP. Proc Natl Acad Sci USA 99:9840-5

Arensbak B, Mikkelsen HB, Gustafsson F, Christensen T, Holstein-Rathlou NH (2001) Expression of connexin 37, 40, and 43 mRNA and protein in renal preglomerular arterioles. Histochem Cell Biol 115:479-87

Bachmann S, Bosse HM, Mundel P (1995) Topography of nitric oxide synthesis by localizing constitutive NO synthases in mammalian kidney. Am J Physiol 268:F885-98

Bader M, Peters J, Baltatu O, Muller DN, Luft FC, Ganten D (2001) Tissue renin-angiotensin systems: new insights from experimental animal models in hypertension research. J Mol Med 79:76-102

Baltatu O, Bader M (2003) Brain renin-angiotensin system. Lessons from functional genomics. Neuroendocrinology 78:253-9

Barajas L, Sokolski KN, Lechago J (1983) Vasoactive intestinal polypeptide-immunoreactive nerves in the kidney. Neurosci Lett 43:263-9

Barrett GL, Morgan TO, Alcorn D (1990) Stimulation of renin synthesis in the hydronephrotic kidney during sodium depletion. Pflugers Arch 415:774-6

Baumbach L, Skott O (1986) Renin release from different parts of rat afferent arterioles in vitro. Am J Physiol 251:F12-6

Beavo JA (1995) Cyclic nucleotide phosphodiesterases: functional implications of multiple isoforms. Physiol Rev 75:725-48

Beierwaltes WH (1995) Selective neuronal nitric oxide synthase inhibition blocks furosemide-stimulated renin secretion in vivo. Am J Physiol 269:F134-9

Beierwaltes WH (1997) Macula densa stimulation of renin is reversed by selective inhibition of neuronal nitric oxide synthase. Am J Physiol 272:R1359-64

Beierwaltes WH (2006) cGMP stimulates renin secretion in vivo by inhibiting phosphodiesterase-3. Am J Physiol Renal Physiol 290:F1376-81

Beierwaltes WH, Potter DL, Shesely EG (2002) Renal baroreceptor-stimulated renin in the eNOS knockout mouse. Am J Physiol Renal Physiol 282:F59-64

Bell PD, Lapointe JY, Sabirov R, Hayashi S, Peti-Peterdi J, Manabe K, Kovacs G, Okada Y (2003) Macula densa cell signaling involves ATP release through a maxi anion channel. Proc Natl Acad Sci USA 100:4322-7

Bie P, Wang BC, Leadley RJ, Jr., Goetz KL (1988) Hemodynamic and renal effects of low-dose infusions of atrial peptide in awakedogs. Am J Physiol 254:R161-9

Bischoff A, Avramidis P, Erdbrugger W, Munter K, Michel MC (1997) Receptor subtypes Y1 and Y5 are involved in the renal effects of neuropeptide Y. Br J Pharmacol 120:1335-43

Blaine EH, Davis JO, Witty RT (1970) Renin release after hemorrhage and after suprarenal aortic constriction in dogs without sodium delivery to the macula densa. Circ Res 27:1081-9

Blaine EH, Davis JO, Prewitt RL (1971) Evidence for a renal vascular receptor in control of renin secretion. Am J Physiol 220:1593-7

Blundell T, Sibanda BL, Pearl L (1983) Three-dimensional structure, specificity and catalytic mechanism of renin. Nature 304:273-5

Bock HA, Hermle M, Brunner FP, Thiel G (1992) Pressure dependent modulation of renin release in isolated perfused glomeruli. Kidney Int 41:275-80

Bolger PM, Eisner GM, Ramwell PW, Slotkoff LM (1976) Effect of prostaglandin synthesis on renal function and renin in the dog. Nature 259:244-5

Bosse HM, Bohm R, Resch S, Bachmann S (1995) Parallel regulation of constitutive NO synthase and renin at JGA of rat kidney under various stimuli. Am J Physiol 269:F793-805
Brands MW, Freeman RH (1988) Aldosterone and renin inhibition by physiological levels of atrial natriuretic factor. Am J Physiol 254:R1011-6
Brands MW, Freeman RH (1989) Aldosterone and renin inhibition by atrial natriuretic factor in potassium-loaded rats. Am J Physiol 257:R1423-8
Brenner BM, Ballermann BJ, Gunning ME, Zeidel ML (1990) Diverse biological actions of atrial natriuretic peptide. Physiol Rev 70:665-99
Breyer MD, Breyer RM (2000) Prostaglandin E receptors and the kidney. Am J Physiol Renal Physiol 279:F12-23
Brown J, O'Flynn MA (1989) Acute effects of physiological increments of alpha-atrial natriuretic peptide in man.Kidney Int 36:645-52
Brown R, Ollerstam A, Johansson B, Skott O, Gebre-Medhin S, Fredholm B, Persson AE (2001) Abolished tubuloglomerular feedback and increased plasma renin in adenosine A1 receptor-deficient mice. Am J Physiol Regul Integr Comp Physiol 281:R1362-7
Brown RD, Thoren P, Steege A, Mrowka R, Sallstrom J, Skott O, Fredholm BB, Persson AE (2006) Influence of the adenosine A1 receptor on blood pressure regulation and renin release. Am J Physiol Regul Integr Comp Physiol 290:R1324-9
Brubacher ES, Vander AJ (1968) Sodium deprivation and renin secretionin unanesthetized dogs. Am J Physiol 214:15-21
Bührle CP, Nobiling R, Taugner R (1985) Intracellular recordings from renin-positive cells of the afferent glomerular arteriole. Am J Physiol 249:F272-81
Bührle CP, Scholz H, Hackenthal E, Nobiling R, Taugner R (1986) Epithelioid cells: membrane potential changes induced by substances influencing renin secretion. Mol Cell Endocrinol 45:37-47
Bunag RD, Page IH, McCubbin JW (1967) Inhibition of renin release by vasopressin and angiotensin. Cardiovasc Res 1:67-73
Burnett JC, Jr., Granger JP, Opgenorth TJ (1984) Effects of synthetic atrial natriuretic factor on renal function and renin release. Am J Physiol 247:F863-6
Butt E, Nolte C, Schulz S, Beltman J, Beavo JA, Jastorff B, Walter U (1992) Analysis of the functional role of cGMP-dependent protein kinase in intact human platelets using a specific activator 8-para-chlorophenylthio-cGMP. Biochem Pharmacol 43:2591-600
Carey RM, McGrath HE, Pentz ES, Gomez RA, Barrett PQ (1997) Biomechanical coupling in renin-releasing cells. J Clin Invest 100:1566-74
Carson P, Carlyle P, Rector TS, Cohn JN (1990) Cardiovascular and neurohormonal effects of atrial natriuretic peptide in conscious dogs with and without chronic left ventricular dysfunction. J Cardiovasc Pharmacol 16:305-11
Castrop H (2007) Mediators of tubuloglomerular feedback regulation of glomerular filtration: ATP and adenosine. Acta Physiol (Oxf)189:3-14
Castrop H, Huang Y, Hashimoto S, Mizel D, Hansen P, Theilig F, Bachmann S, Deng C, Briggs J, Schnermann J (2004a) Impairment of tubuloglomerular feedback regulation of GFR in ecto-5'-nucleotidase/CD73-deficient mice. J Clin Invest 114:634-42
Castrop H, Schweda F, Mizel D, Huang Y, Briggs J, Kurtz A, Schnermann J (2004b) Permissive role of nitric oxide in macula densa control of renin secretion. Am J Physiol Renal Physiol 286:F848-57
Castrop H, Lorenz JN, Hansen PB, Friis U, Mizel D, Oppermann M, Jensen BL, Briggs J, Skott O, Schnermann J (2005) Contribution of the basolateral isoform of the Na-K-2Cl- cotransporter (NKCC1/BSC2) to renin secretion. Am J Physiol Renal Physiol 289:F1185-92
Chen M, Harris MP, Rose D, Smart A, He XR, Kretzler M, Briggs JP, Schnermann J (1994) Renin and renin mRNA in proximal tubules of therat kidney. J Clin Invest 94:237-43
Chen L, Kim SM, Oppermann M, Faulhaber-Walter R, Huang Y, Mizel D, Chen M, Lopez ML,Weinstein LS, Gomez RA, Briggs JP, Schnermann J (2007) Regulation of renin in mice with Cre recombinase-mediated deletion of G protein Gsalpha injuxtaglomerular cells. Am J Physiol Renal Physiol 292:F27-37

Cheng HF, Wang JL, Zhang MZ, Miyazaki Y, Ichikawa I, McKanna JA, Harris RC (1999) Angiotensin II attenuates renal cortical cyclooxygenase-2 expression. J Clin Invest 103:953-61

Chiu N, Park I, Reid IA (1996) Stimulation of renin secretion by the phosphodiesterase IV inhibitor rolipram. J Pharmacol Exp Ther 276:1073-7

Chiu T, Reid IA (1996) Role of cyclic GMP-inhibitable phosphodiesterase and nitric oxide in the beta adrenoceptor control of renin secretion. J Pharmacol Exp Ther 278:793-9

Chiu YJ, Hu SH, Reid IA (1999) Inhibition of phosphodiesterase III with milrinone increases renin secretion in human subjects. J Pharmacol Exp Ther 290:16-9

Churchill PC (1980) Effect of D-600 on inhibition of in vitro renin release in the rat by high extracellular potassium and angiotensin II. J Physiol 304:449-58

Churchill PC (1985) Second messengers in renin secretion. Am J Physiol 249:F175-84

Churchill PC, Churchill MC (1985) A1 and A2 adenosine receptor activation inhibits and stimulates renin secretion of rat renal cortical slices. J Pharmacol Exp Ther 232:589-94

Clark AF, Sharp MG, Morley SD, Fleming S, Peters J, Mullins JJ (1997) Renin-1 is essential for normal renal juxtaglomerular cell granulation and macula densa morphology. J Biol Chem 272:18185-90

Cohen Y, Rahamimov R, Naveh-Many T, Silver J, Rahamimoff R (1997) Where is the "inverting factor" in hormone secretion from parathyroid cells? Am J Physiol 273:E631-7

Cotrina ML, Lin JH, Alves-Rodrigues A, Liu S, Li J, Azmi-Ghadimi H, Kang J, Naus CC, Nedergaard M (1998) Connexins regulate calcium signaling by controlling ATP release. Proc Natl Acad Sci USA 95:15735-40

Crackower MA, Sarao R, Oudit GY, Yagil C, Kozieradzki I, Scanga SE, Oliveira-dos-Santos AJ, da Costa J, Zhang L, Pei Y, Scholey J, Ferrario CM, Manoukian AS, Chappell MC, Backx PH, Yagil Y, Penninger JM (2002) Angiotensin-converting enzyme 2 is an essential regulator of heart function. Nature 417:822-8

Crowley SD, Gurley SB, Oliverio MI, Pazmino AK, Griffiths R, Flannery PJ, Spurney RF, Kim HS, Smithies O, Le TH, Coffman TM (2005) Distinct roles for the kidney and systemic tissues in blood pressure regulation by the renin-angiotensin system. J Clin Invest 115:1092-9

Cuneo RC, Espiner EA, Nicholls MG, Yandle TG, Livesey JH (1987) Effect of physiological levels of atrial natriuretic peptide on hormone secretion: inhibition of angiotensin-induced aldosterone secretion and renin release in normal man. J Clin Endocrinol Metab 65:765-72

Danser AH (2003) Local renin-angiotensin systems: the unanswered questions. Int J Biochem Cell Biol 35:759-68

Danser AH, Batenburg WW, van Esch JH (2007) Prorenin and the (pro)renin receptor—an update. Nephrol Dial Transplant 22:1288-92

de Wit C, Roos F, Bolz SS, Pohl U (2003) Lack of vascular connexin 40 is associated with hypertension and irregular arteriolar vasomotion. Physiol Genomics 13:169-77

Deray G, Branch RA, Herzer WA, Ohnishi A, Jackson EK (1987) Effects of atrial natriuretic factor on hormone-induced renin release. Hypertension 9:513-7

DiBona GF, Kopp UC (1997) Neural control of renal function. Physiol Rev 77:75-197

DiBona GF, Sawin LL (1985) Renal nerve activity in conscious rats during volume expansion and depletion. Am J Physiol 248:F15-23

Dickinson DP, Gross KW, Piccini N, Wilson CM (1984) Evolution and variation of renin genes in mice. Genetics108:651-67

Dzau VJ (1993) Local expression and pathophysiological role of renin-angiotensin in the blood vessels and heart. Basic Res Cardiol 88 Suppl 1:1-14

DzauVJ, Ellison KE, Brody T, Ingelfinger J, Pratt RE (1987) A comparative study of the distributions of renin and angiotensinogen messenger ribonucleic acids in rat and mouse tissues. Endocrinology 120:2334-8

Ehmke H, Persson PB, Just A, Nafz B, Seyfarth M, Hackenthal E, Kirchheim HR (1992) Physiological concentrations of ANP exert a dual regulatory influence on renin release in conscious dogs. Am J Physiol 263:R529-36

Ekker M, Tronik D, Rougeon F(1989) Extra-renal transcription of the renin genes in multiple tissues of mice and rats. Proc Natl Acad Sci USA 86:5155-8

Evans WH, De Vuyst E, Leybaert L (2006) The gap junction cellular internet: connexin hemichannels enter the signalling limelight. Biochem J3 97:1-14
Fan L, Archambault D, Chavez S, Aljabari AW, Quillen EW, Jr. (1994a) Influences of renal nerves and pregnancy on renin secretion in sheep. Am J Physiol 266:R668-73
Fan L, Mukaddam-Daher S, Gutkowska J, Nuwayhid BS, Quillen EW, Jr. (1994b) Renal perfusion pressure and renin secretion in bilaterally renal denervated sheep. Can J Physiol Pharmacol 72:782-7
Fernandez-Repollet E, Silva-Netto CR, Colindres RE, Gottschalk CW (1985) Role of renal nerves in maintaining sodium balance in unrestrained conscious rats. Am J Physiol 249:F819-26
Ferrier CP, Kurtz A, Lehner P, Shaw SG, Pusterla C, Saxenhofer H, Weidmann P (1989) Stimulation of renin secretion by potassium-channel activation with cromakalim. Eur J Clin Pharmacol 36:443-7
Franco-Saenz R, Suzuki S, Tan SY,Mulrow PJ (1980) Prostaglandin stimulation of renin release: independence of beta-adrenergic receptor activity and possible mechanism of action. Endocrinology 106:1400-4
Freestone S, MacDonald TM, Jeffrey RF, Brown J, Lee MR (1989) The renal effects of atrial natriuretic peptide in man are not attenuated by (+)-sulpiride. Br J Clin Pharmacol 27:13-8
Friberg P, Meredith I, Jennings G, Lambert G, Fazio V, Esler M (1990) Evidence for increased renal norepinephrine overflow during sodium restriction in humans. Hypertension 16:121-30
Friis UG, Jensen BL, Aas JK, Skott O (1999) Direct demonstration of exocytosis and endocytosis in single mouse juxtaglomerular cells. Circ Res 84:929-36
Friis UG, Jensen BL, Sethi S, Andreasen D, Hansen PB, Skott O (2002) Control of renin secretion from rat juxtaglomerular cells by cAMP-specific phosphodiesterases. Circ Res 90:996-1003
Friis UG, Jorgensen F, Andreasen D, Jensen BL, Skott O (2003) Molecular and functional identification of cyclic AMP-sensitive BKC a potassium channels (ZERO variant) and L-type voltage-dependent calcium channels in single rat juxtaglomerular cells. Circ Res 93:213-20
Friis UG, Jorgensen F, Andreasen D, Jensen BL, Skott O (2004) Membrane potential and cation channels in rat juxtaglomerular cells. Acta Physiol Scand 181:391-6
Friis UG, Stubbe J, Uhrenholt TR, Svenningsen P, Nusing RM, Skott O, Jensen BL (2005) Prostaglandin E_2 EP2 and EP4 receptor activation mediates cAMP-dependent hyperpolarization and exocytosis of renin in juxtaglomerular cells. Am J Physiol Renal Physiol 289:F989-97
Fujino T, Nakagawa N, Yuhki K, Hara A, Yamada T, Takayama K, Kuriyama S, Hosoki Y, Takahata O, Taniguchi T, Fukuzawa J, Hasebe N, Kikuchi K, Narumiya S, Ushikubi F (2004) Decreased susceptibility to renovascular hypertension in mice lacking the prostaglandin I_2 receptor IP. J Clin Invest 114:805-12
Gambaryan S, Hausler C, Markert T, Pohler D, Jarchau T, Walter U, Haase W, Kurtz A, Lohmann SM (1996) Expression of type II cGMP-dependent protein kinase in rat kidney is regulated by dehydration and correlated with renin gene expression. J Clin Invest 98:662-70
Gambaryan S, Wagner C, Smolenski A,Walter U, Poller W, Haase W, Kurtz A, Lohmann SM (1998) Endogenous or overexpressed cGMP-dependent protein kinases inhibit cAMP-dependent renin release from rat isolated perfused kidney, microdissected glomeruli, and isolated juxtaglomerular cells. Proc Natl Acad Sci USA 95:9003-8
Ganten D, Minnich JL, Granger P, Hayduk K, Brecht HM, Barbeau A, Boucher R, Genest J (1971) Angiotensin-forming enzyme in brain tissue. Science 173:64-5
Garcia R, Thibault G, Gutkowska J, Hamet P, Cantin M, Genest J (1985) Effect of chronic infusion of synthetic atrial natriuretic factor (ANF 8-33) in conscious two-kidney, one-clip hypertensive rats. Proc Soc Exp Biol Med 178:155-9
Gerber JG, Branch RA, Nies AS, Hollifield JW, Gerkens JF (1979a) Influence of hypertonic saline on canine renal blood flow and renin release. Am J Physiol 237:F441-6
Gerber JG, Keller RT, Nies AS (1979b) Prostaglandins and renin release: the effect of PGI_2, PGE_2, and 13,14-dihydro PGE_2 on the baroreceptor mechanism of renin release in the dog. Circ Res 44:796-9
Ginesi LM, Munday KA, Noble AR (1983) Secretion control for active and inactive renin: effects of calcium and potassium on rabbit kidney cortex slices. J Physiol 344:453-63

Gnaedinger MP, Uehlinger DE, Weidmann P, Sha SG, Muff R,Born W, Rascher W, Fischer JA (1989) Distinct hemodynamic and renal effects of calcitonin gene-related peptide and calcitonin in men. Am J Physiol 257:E848-54

Grandis DJ, Uretsky BF, Verbalis JG, Chong TK, Puschett JB (1992) Relationship of the renin-angiotensin system and systemic arterial pressure to sodium excretion during atrial natriuretic peptide infusion in men. Am J Hypertens 5:793-9

Greenberg SG, He XR, Schnermann JB, Briggs JP (1995) Effect of nitric oxide on renin secretion. I. Studies in isolated juxtaglomerular granular cells. Am J Physiol 268:F948-52

Grünberger C, Obermayer B, Klar J, Kurtz A, Schweda F (2006) The calcium paradoxon of renin release: calcium suppresses renin exocytosis by inhibition of calcium-dependent adenylate cyclases AC5 and AC6.Circ Res 99:1197-206

Hackenthal E, Schwertschlag U, Seyberth HW (1980) Prostaglandins and renin release studies in the isolated perfused rat kidney. Prog Biochem Pharmacol 17:98-107

Hackenthal E, Aktories K, Jakobs KH, Lang RE (1987) Neuropeptide Y inhibits renin release by a pertussis toxin-sensitive mechanism. Am J Physiol 252:F543-50

Hackenthal E, Paul M, Ganten D, Taugner R (1990) Morphology, physiology, and molecular biology of renin secretion. Physiol Rev 70:1067-116

Haefliger JA, Demotz S, Braissant O, Suter E, Waeber B, Nicod P, Meda P (2001) Connexins 40 and 43 are differentially regulated within the kidneys of rats with renovascular hypertension. Kidney Int 60:190-201

Haefliger JA, Krattinger N, Martin D, Pedrazzini T, Capponi A, Doring B, Plum A, Charollais A, Willecke K, Meda P (2006) Connexin43-dependent mechanism modulates renin secretion and hypertension. J Clin Invest 116:405-13

Hannaert P, Alvarez-Guerra M, Pirot D, Nazaret C, Garay RP (2002) Rat NKCC2/NKCC1 cotransporter selectivity for loop diuretic drugs. Naunyn Schmiedebergs Arch Pharmacol 365:193-9

Hanner F, Chambrey R, Bourgeois S, Meer E, Mucsi I, Rosivall L, Shull GE, Lorenz JN, Eladari D, Peti-Peterdi J (2008) Increased renal renin content in mice lacking the Na^+/H^+ exchanger NHE2. Am J Physiol Renal Physiol 294:F937-44

Hao CM, Breyer MD (2007) Physiologic and pathophysiologic roles of lipid mediators in the kidney. Kidney Int 71:1105-15

Harding P, Sigmon DH, Alfie ME, Huang PL, Fishman MC, Beierwaltes WH, Carretero OA (1997) Cyclooxygenase-2 mediates increased renal renin content induced by low-sodium diet. Hypertension 29:297-302

Hardman JA, Hort YJ, Catanzaro DF, Tellam JT, Baxter JD, Morris BJ, Shine J (1984) Primary structure of the human renin gene. DNA 3:457-68

Harris RC, McKanna JA, Akai Y, Jacobson HR, Dubois RN, Breyer MD (1994) Cyclooxygenase-2 is associated with the macula densa of rat kidney and increases with salt restriction. J Clin Invest 94:2504-10

Hartner A, Goppelt-Struebe M, Hilgers KF (1998) Coordinate expression of cyclooxygenase-2 and renin in the rat kidney in renovascular hypertension. Hypertension 31:201-5

Hartner A, Cordasic N, Goppelt-Struebe M, Veelken R, Hilgers KF (2003) Role of macula densa-cyclooxygenase-2 in renovascular hypertension. Am J Physiol Renal Physiol 284:F498-502

Hautmann M, Friis UG, Desch M, Todorov V, Castrop H, Segerer F, Otto C, Schutz G, Schweda F (2007) Pituitary adenylate cyclase-activating polypeptide stimulates renin secretion via activation of PAC1 receptors. J Am Soc Nephrol 18:1150-6

He XR, Greenberg SG, Briggs JP, Schnermann J (1995) Effects of furosemide and verapamil on the NaCl dependency of maculadensa-mediated renin secretion. Hypertension 26:137-42

Henrich WL, Campbell WB (1984) Relationship between PG and beta-adrenergic pathways to renin release in rat renal cortical slices. AmJ Physiol 247:E343-8

Henrich WL, Needleman P, Campbell WB (1986) Effect of atriopeptin III on renin release in vitro. Life Sci 39:993-1001

Henrich WL, McAlister EA, Smith PB, Lipton J, Campbell WB (1987) Direct inhibitory effect of atriopeptin III on renin release in primate kidney. Life Sci 41:259-64

Henrich WL, McAllister EA, Smith PB, Campbell WB (1988) Guanosine 3',5'-cyclic monophosphate as a mediator of inhibition of renin release. Am J Physiol 255:F474-8

Hirata Y, Ishii M, Sugimoto T, Matsuoka H, Ishimitsu T, Atarashi K, Miyata A, Kangawa K, Matsuo H (1987) Relationship between the renin-aldosterone system and atrial natriuretic polypeptide in rats. Clin Sci (Lond) 72:165-70

Hirota N, Ichihara A, Koura Y, Hayashi M, Saruta T (2002) Phospholipase D contributes to transmural pressure control of proren in processing in juxtaglomerular cell. Hypertension 39:363-7

Hiruma M, Ikemoto F, Yamamoto K (1986) Rat atrial natriuretic factor stimulates renin release from renal cortical slices. Eur J Pharmacol 125:151-3

Höcherl K, Kammerl M, Kees F, Kramer BK, Grobecker HF, Kurtz A (2002a) Role of renal nerves in stimulation of renin, COX-2, and nNOS in rat renal cortex during salt deficiency. Am J Physiol Renal Physiol 282:F478-84

Höcherl K, Kammerl MC, Schumacher K, Endemann D, Grobecker HF, Kurtz A (2002b) Role of prostanoids in regulation of therenin-angiotensin-aldosterone system by salt intake. Am J Physiol Renal Physiol 283:F294-301

Holdaas H, DiBona GF, Kiil F(1981) Effect of low-level renal nerve stimulation on renin release from nonfiltering kidneys. Am J Physiol 241:F156-61

Holmer S, Eckardt KU, LeHir M, Schricker K, Riegger G, Kurtz A (1993) Influence of dietary NaCl intake on renin gene expression in the kidneys and adrenal glands of rats. Pflugers Arch 425:62-7

Holtwick R, Baba HA, Ehler E, Risse D, Vobeta M, Gehrmann J, Pierkes M, Kuhn M (2002) Left but not right cardiac hypertrophy in atrial natriuretic peptide receptor-deficient mice is prevented by angiotensin type 1receptor antagonist losartan. J Cardiovasc Pharmacol 40:725-34

Hwan Seul K, Beyer EC (2000) Heterogeneous localization of connexin40 in the renal vasculature. Microvasc Res 59:140-8

Ichihara A, Suzuki H, Murakami M, Naitoh M, Matsumoto A, Saruta T (1995) Interactions between angiotensin II and norepinephrine on renin release by juxtaglomerular cells. Eur J Endocrinol 133:569-77

Ichihara A, Suzuki H, Miyashita Y, Naitoh M, Hayashi M, Saruta T (1999) Transmural pressure inhibits prorenin processing injuxtaglomerular cell. Am J Physiol 277:R220-8

Ichihara A, Kobori H, Nishiyama A, Navar LG (2004) Renal renin-angiotensin system. Contrib Nephrol 143:117-30

Imanishi M, Tsuji T, Nakamura S, Takamiya M (2001) Prostaglandin I_2/E_2 ratios in unilateral renovascular hypertension of different severities. Hypertension 38:23-9

Imbs JL, Schmidt M, Schwartz J (1975) Effect of dopamineon renin secretion in the anesthetized dog. Eur J Pharmacol 33:151-7

Ishimitsu T, Hirata Y, Matsuoka H, Ishii M, Sugimoto T, Kangawa K, Matsuo H (1992) In vivo and in vitro effects ofatrial natriuretic peptide on renin release. Clin Exp Pharmacol Physiol 19:711-6

Itskovitz J, Bruneval P, Soubrier F, Thaler I, Corvol P, Sealey JE (1992) Localization of renin gene expression to monkey ovarian theca cells by in situ hybridization. J Clin Endocrinol Metab 75:1374-80

Ito S, Carretero OA, Abe K, Beierwaltes WH, Yoshinaga K (1989) Effect of prostanoids on renin release from rabbit afferent arterioles with and without macula densa. Kidney Int 35:1138-44

Itoh S, Carretero OA (1985) Role of the macula densa in renin release. Hypertension 7:I49-54

Itoh S, Abe K, Nushiro N, Omata K, Yasujima M, Yoshinaga K (1987) Effect of atrial natriuretic factor on renin release in isolated afferent arterioles. Kidney Int 32:493-7

Jackson EK, Gerkens JF, Brash AR, Branch RA (1982) Acute renal artery constriction increases renal prostaglandin I_2 biosynthesis and renin release in the conscious dog. J Pharmacol Exp Ther 222:410-3

Jahn R, Lang T, Sudhof TC (2003) Membrane fusion. Cell 112:519-33

Jensen BL,Skott O (1996) Blockade of chloride channels by DIDS stimulates renin release and inhibits contraction of afferent arterioles. Am J Physiol 270:F718-27

Jensen BL, Schmid C, Kurtz A(1996) Prostaglandins stimulate renin secretion and renin mRNA in mouse renal juxtaglomerular cells. Am J Physiol 271:F659-69

Jensen BL, Kramer BK, Kurtz A (1997) Adrenomedullin stimulates renin release and renin mRNA in mouse juxtaglomerular granular cells. Hypertension 29:1148-55

Jensen BL, Gambaryan S, Scholz H, Kurtz A (1998) KATP channels are not essential for pressure-dependent control of renin secretion. Pflugers Arch 435:670-7

Jutras I, Reudelhuber TL (1999) Prorenin processing by cathepsin B in vitro and in transfected cells. FEBS Lett 443:48-52

Kageyama S, Brown J (1990) Effect of atrial natriuretic peptide on renin release in rat isolated glomeruli. Biochem Biophys Res Commun 168:37-42

Kammerl MC, Nusing RM, Richthammer W, Kramer BK, Kurtz A (2001a) Inhibition of COX-2 counteracts the effects of diuretics in rats. Kidney Int 60:1684-91

Kammerl MC, Nusing RM, Schweda F, Endemann D, Stubanus M,Kees F, Lackner KJ, Fischereder M, Kramer BK (2001b) Low sodium andfurosemide-induced stimulation of the renin system in man is mediated bycyclooxygenase 2. Clin Pharmacol Ther 70:468-74

Kaplan MR, Plotkin MD, Brown D, Hebert SC, Delpire E (1996) Expression of the mouse Na-K-2Cl cotransporter, mBSC2, in the terminal inner medullary collecting duct, the glomerular and extraglomerular mesangium, and the glomerular afferent arteriole. J Clin Invest 98:723-30

Keeton TK, Campbell WB (1980) The pharmacologicalteration of renin release. Pharmacol Rev 32:81-227

Kikkawa Y, Yamanaka N, Tada J, Kanamori N, Tsumura K,Hosoi K (1998) Prorenin processing and restricted endoproteolysis by mousetissue kallikrein family enzymes (mK1, mK9, mK13, and mK22). Biochim BiophysActa 1382:55-64

Kim SM, Mizel D, Huang YG, Briggs JP, Schnermann J (2006) Adenosine as a mediator of macula densa-dependent inhibition of renin secretion. Am J Physiol Renal Physiol 290:F1016-23

Kim SM, Chen L, Faulhaber-Walter R, Oppermann M, Huang Y, Mizel D, Briggs JP, Schnermann J (2007a) Regulation of renin secretion and expression in mice deficient in beta1- and beta2-adrenergicreceptors. Hypertension 50:103-9

Kim SM, Chen L, Mizel D, Huang YG, Briggs JP, Schnermann J (2007b) Low plasma renin and reduced renin secretory responses to acute stimuli in conscious COX-2-deficient mice. Am J Physiol Renal Physiol 292:F415-22

King JA, Lush DJ, Fray JC (1993) Regulation of renin processing and secretion: chemiosmotic control and novel secretory pathway. Am J Physiol 265:C305-20

Kirchheim HR, Ehmke H, Hackenthal E, Lowe W, Persson P (1987) Autoregulation of renal blood flow, glomerular filtration rate and renin release in conscious dogs. Pflugers Arch 410:441-9

Kirchner KA, Kotchen TA, Galla JH, Luke RG (1978) Importance of chloride for acute inhibition of renin by sodium chloride. Am J Physiol 235:F444-50

Klar J, Sandner P, Muller MW, Kurtz A (2002) Cyclic AMP stimulates renin gene transcription in juxtaglomerular cells. Pflugers Arch 444:335-44

Knoblich PR, Freeman RH, Villarreal D (1996) Pressure-dependent renin release during chronic blockade of nitric oxidesynthase. Hypertension 28:738-42

Kobori H, Nangaku M, Navar LG, Nishiyama A (2007) The intrarenal renin-angiotensinsystem: from physiology to the pathobiology of hypertension and kidney disease. Pharmacol Rev 59:251-87

Komlosi P, Peti-Peterdi J, Fuson AL, Fintha A, Rosivall L, Bell PD (2004) Macula densabasolateral ATP release is regulated by luminal [NaCl] and dietary salt intake. Am J Physiol Renal Physiol 286:F1054-8

Kompanowska-Jezierska E, Wolff H, Kuczeriszka M, Gramsbergen JB,Walkowska A, Johns EJ, Bie P (2008) Renal nerves and nNOS: roles in natriuresisof acute isovolumetric sodium loading in conscious rats. Am J Physiol RegulIntegr Comp Physiol 294:R1130-9

Krattinger N, Capponi A, Mazzolai L, Aubert JF, Caille D, Nicod P, Waeber G, Meda P, Haefliger JA (2007) Connexin 40 regulates renin production and blood pressure. Kidney Int 72:814-22

Kuhn M (2005) Cardiac and intestinal natriuretic peptides: insights from genetically modified mice. Peptides 26:1078-85

Kurtz A (1986) Transmembrane signalling of atrial natriuretic peptide in rat renal juxtaglomerular cells. Klin Wochenschr 64 Suppl 6:37-41
Kurtz A (1989) Cellular control of renin secretion. Rev Physiol Biochem Pharmacol 113:1-40
Kurtz A, Penner R (1989) Angiotensin II induces oscillations of intracellular calcium and blocks anomalous inward rectifying potassium current in mouse renal juxtaglomerular cells. Proc Natl Acad Sci USA 86:3423-7
Kurtz A, Schweda F (2006) Osmolarity-induced renin secretion from kidneys: evidence for readily releasable renin pools. Am J Physiol Renal Physiol 290:F797-805
Kurtz A, Wagner C (1998) Role of nitric oxide in the control of renin secretion. Am J Physiol 275:F849-62
Kurtz A, Wagner C (1999) Cellular control of renin secretion. J Exp Biol 202:219-25
Kurtz A, Pfeilschifter J, Bauer C (1984) Is renin secretion governed by the calcium permeability of the juxtaglomerular cell membrane? Biochem Biophys Res Commun 124:359-66
Kurtz A, Della Bruna R, Pfeilschifter J, Bauer C (1986a) Effect of synthetic atrial natriuretic peptide on rat renal juxtaglomerular cells. J Hypertens Suppl 4:S57-60
Kurtz A, Della Bruna R, Pfeilschifter J, Taugner R, Bauer C (1986b) Atrial natriuretic peptide inhibits renin release from juxtaglomerular cells by a cGMP-mediated process. Proc Natl Acad Sci USA 83:4769-73
Kurtz A, Pfeilschifter J, Hutter A, Buhrle C, Nobiling R, Taugner R, Hackenthal E, Bauer C (1986c) Role of protein kinase C in inhibition of renin release caused by vasoconstrictors. Am J Physiol 250:C563-71
Kurtz A, Della Bruna R, Pfeilschifter J, Bauer C (1988a) Role of cGMP as second messenger of adenosine in the inhibition of renin release. Kidney Int 33:798-803
Kurtz A, Della Bruna R, Pratz J, Cavero I (1988b) Rat juxtaglomerular cells are endowed with DA-1 dopamine receptors mediating renin release. J Cardiovasc Pharmacol 12:658-63
Kurtz A, Muff R, Born W, Lundberg JM, Millberg BI, Gnadinger MP, Uehlinger DE, Weidmann P, Hokfelt T, Fischer JA (1988c) Calcitonin gene-related peptide is a stimulator of renin secretion. J Clin Invest 82:538-43
Kurtz A, Skott O, Chegini S, Penner R (1990) Lack of direct evidence for a functional role of voltage-operated calcium channels in juxtaglomerular cells. Pflugers Arch 416:281-7
Kurtz A, Kaissling B, Busse R, Baier W (1991) Endothelial cells modulate renin secretion from isolated mouse juxtaglomerular cells. J Clin Invest 88:1147-54
Kurtz A, Gotz KH, Hamann M, Kieninger M, Wagner C (1998a) Stimulation of renin secretion by NO donors is related to the cAMP pathway. Am J Physiol 274:F709-17
Kurtz A, Gotz KH, Hamann M, Wagner C (1998b) Stimulation of renin secretion by nitric oxide is mediated by phosphodiesterase 3. Proc Natl Acad Sci U S A 95:4743-7
Kurtz A, Hamann M, Götz K (2000) Role of potassium channels in the control of renin secretion from isolated perfused rat kidneys. Pflugers Arch 440:889-95
Kurtz L, Schweda F, de Wit C, Kriz W, Witzgall R, Warth R, Sauter A, Kurtz A, Wagner C (2007) Lack of connexin 40 causes displacement of renin-producing cells from afferent arterioles to the extraglomerular mesangium. J Am Soc Nephrol18:1103-11
Laframboise M, Reudelhuber TL, Jutras I, Brechler V, Seidah NG, Day R, Gross KW, Deschepper CF (1997) Prorenin activation and prohormone convertases in the mouse As4.1 cell line. Kidney Int51:104-9
Leichtle A, Rauch U, Albinus M, Benohr P, Kalbacher H, Mack AF, Veh RW, Quast U, Russ U (2004) Electrophysiological and molecular characterization of the inward rectifier injuxtaglomerular cells from rat kidney. J Physiol 560:365-76
Leung PS (2007) The physiology of a local renin-angiotensin system in the pancreas. J Physiol 580:31-7
Leung PS, Sernia C (2003) The renin-angiotensin system and male reproduction: new functions for old hormones. J Mol Endocrino l30:263-70
Lorenz JN, Kotchen TA, Ott CE (1990) Effect of Na and Cl infusion on loop function and plasma renin activity in rats. Am J Physiol 258:F1328-35

Lorenz JN, Weihprecht H, Schnermann J, Skott O, Briggs JP (1991) Renin release from isolated juxtaglomerular apparatus depends on macula densa chloride transport. Am J Physiol 260:F486-93

Loutzenhiser R, Chilton L, Trottier G (1997) Membrane potential measurements in renal afferent and efferent arterioles: actions of angiotensin II. Am J Physiol 273:F307-14

Maack T, Marion DN, Camargo MJ, Kleinert HD, Laragh JH, Vaughan ED, Jr., Atlas SA (1984) Effects of auriculin (atrial natriuretic factor) on blood pressure, renal function, and therenin-aldosterone system in dogs. Am J Med 77:1069-75

Mackins CJ, Kano S, Seyedi N, Schafer U, Reid AC, Machida T, Silver RB, Levi R (2006) Cardiac mast cell-derived renin promotes local angiotensin formation, norepinephrine release, and arrhythmias inischemia/reperfusion. J Clin Invest 116:1063-70

Mann B, Hartner A, Jensen BL, Hilgers KF, Hocherl K, Kramer BK, Kurtz A (2001a) Acute upregulation of COX-2 by renal arterystenosis. Am J Physiol Renal Physiol 280:F119-25

Mann B, Hartner A, Jensen BL, Kammerl M, Kramer BK, Kurtz A (2001b) Furosemide stimulates macula densa cyclooxygenase-2 expression in rats. Kidney Int 59:62-8

Matsuba H, Watanabe T, Watanabe M, Ishii Y, Waguri S, Kominami E, Uchiyama Y (1989) Immunocytochemical localization of prorenin, renin, and cathepsins B, H, and Lin juxta-glomerular cells of rat kidney. J Histochem Cytochem 37:1689-97

Matsusaka T, Nishimura H, Utsunomiya H, Kakuchi J, Niimura F, Inagami T, Fogo A, Ichikawa I (1996) Chimeric mice carrying 'regional' targeted deletion of the angiotensin type 1Areceptor gene. Evidence against the role for local angiotensin in the in vivo feedback regulation of renin synthesis in juxtaglomerular cells. J Clin Invest 98:1867-77

Matzdorf C, Kurtz A, Hocherl K (2007) COX-2 activity determines the level of renin expression but is dispensable for acute upregulation of renin expression in rat kidneys. Am J Physiol Renal Physiol 292:F1782-90

MeloLG,VeressAT, ChongCK, PangSC,FlynnTG,SonnenbergH(1998)Salt-sensitivehypertension in ANP knockout mice: potential role of abnormal plasma renin activity. Am J Physiol 274:R255-61

Melo LG, Steinhelper ME, Pang SC, Tse Y, Ackermann U (2000) ANP in regulation of arterial pressure and fluid-electrolyte balance: lessons from genetic mouse models. Physiol Genomics 3:45-58

Mercure C, Jutras I, Day R, Seidah NG, Reudelhuber TL (1996) Prohormone convertase PC5 is a candidate processing enzyme for prorenin in the human adrenal cortex. Hypertension 28:840-6

Michelakis AM (1971)The effect of angiotensin on renin production and release in vitro. Proc Soc Exp Biol Med 138:1106-8

Moe O, Tejedor A, Campbell WB, Alpern RJ, Henrich WL (1991) Effects of endothelin on in vitro renin secretion. Am J Physiol 260:E521-5

Moe OW, Ujiie K, Star RA, Miller RT, Widell J, Alpern RJ, Henrich WL (1993) Renin expression in renal proximal tubule. J Clin Invest 91:774-9

Moore PK, Bland-Ward PA (1996) 7-Nitroindazole: an inhibitor of nitric oxide synthase. Methods Enzymol 268:393-8

Mundel P, Bachmann S, Bader M, Fischer A, Kummer W, Mayer B, Kriz W (1992) Expression of nitric oxide synthase in kidney macula densa cells. Kidney Int 42:1017-9

Nabel C, Schweda F, Riegger GA, Kramer BK, Kurtz A (1999) Chloride channel blockers attenuate the inhibition of renin secretion by angiotensin II. Pflugers Arch 438:694-9

Narumi S, Yasui T, Yoshizawa M, Kawamura M, Suzuki H, Nakane H, Saruta T (1987) Effects of atrial natriuretic peptide on renal function and renin release in the isolated perfused rat kidney. Jpn Heart J 28:221-7

Nausch LW, Ledoux J, Bonev AD, Nelson MT, Dostmann WR (2008) Differential patterning of cGMP in vascular smooth muscle cells revealed by single GFP-linked biosensors. Proc Natl Acad Sci USA 105:365-70

Neher E, Marty A (1982) Discrete changes of cell membrane capacitance observed under conditions of enhanced secretion in bovine adrenal chromaffin cells. Proc Natl Acad Sci USA 79:6712-6

Neves FA, Duncan KG, Baxter JD (1996) Cathepsin B is a prorenin processing enzyme. Hypertension 27:514-7
Nguyen G (2006) Renin/prorenin receptors. Kidney Int 69:1503-6
Noble AR, Abu-Kishk RA, D'Aloia MA, Williams BC, Lush DJ (1994) Cyclic GMP-linked pathway for renin secretion. Kidney Int 46:1588-90
O'Tierney PF, Komolova M, Tse MY, Adams MA, Pang SC (2008) Altered regulation of renal interstitial hydrostatic pressure and the renal renin-angiotensin system in the absence of atrial natriuretic peptide. J Hypertens 26:303-11
Obana K, Naruse M, Naruse K, Sakurai H, Demura H, Inagami T, Shizume K (1985) Synthetic rat atrial natriuretic factor inhibits in vitro and in vivo renin secretion in rats. Endocrinology 117:1282-4
Ogawa K, Yamasato M, Taniguchi K (1995) Exocytosis of secretory granules in the juxtaglomerular granular cells of kidneys. Anat Rec 243:336-46
Oliver JA, Pinto J, Sciacca RR, Cannon PJ (1980) Increased renal secretion of norepinephrine and prostaglandin E_2 during sodium depletion in the dog. J Clin Invest 66:748-56
Onda T, Hashimoto Y, Nagai M, Kuramochi H, Saito S, Yamazaki H, Toya Y, Sakai I, Homcy CJ, Nishikawa K, Ishikawa Y (2001) Type-specific regulation of adenylyl cyclase. Selective pharmacological stimulation and inhibition of adenylyl cyclase isoforms. J Biol Chem 276:47785-93
Opgenorth TJ, Burnett JC, Jr., Granger JP, Scriven TA (1986) Effects of atrial natriuretic peptide on renin secretion in nonfiltering kidney. Am J Physiol 250:F798-801
Oppermann M, Mizel D, Huang G, Li C, Deng C, Theilig F, Bachmann S, Briggs J, Schnermann J, Castrop H (2006) Macula densa control of renin secretion and preglomerular resistance in mice with selective deletion of the B isoform of the Na,K,2Cl co-transporter. J Am Soc Nephrol 17:2143-52
Oppermann M, Mizel D, Kim SM, Chen L, Faulhaber-Walter R, Huang Y, Li C, Deng C, Briggs J, Schnermann J, Castrop H (2007) Renal function in mice with targeted disruption of the A isoform of the Na-K-2Cl co-transporter. J Am Soc Nephrol 18:440-8
Oppermann M, Friedman DJ, Faulhaber-Walter R, Mizel D, Castrop H, Enjyoji K, Robson SC, Schnermann J (2008) Tubuloglomerular feedback and renin secretion in NTPDase1/CD39-deficient mice. Am J Physiol Renal Physiol 294:F965-70
Ortiz-Capisano MC, Ortiz PA, Garvin JL, Harding P, Beierwaltes WH (2007a) Expression and function of the calcium-sensing receptor in juxtaglomerular cells. Hypertension 50:737-43
Ortiz-Capisano MC, Ortiz PA, Harding P, Garvin JL, Beierwaltes WH (2007b) Adenylyl cyclase isoform v mediates renin release from juxtaglomerular cells. Hypertension 49:618-24
Ortiz-Capisano MC, Ortiz PA, Harding P, Garvin JL, Beierwaltes WH (2007c) Decreased intracellular calcium stimulates renin release via calcium-inhibitable adenylyl cyclase. Hypertension 49:162-9
Osswald H, Nabakowski G, Hermes H (1980) Adenosine as a possible mediator of metabolic control of glomerular filtration rate. Int J Biochem 12:263-7
Pandey KN, Maki M, Inagami T (1984) Detection of renin mRNA in mouse testis by hybridization with renin cDNA probe. Biochem Biophys Res Commun 125:662-7
Paul M, Wagner D, Metzger R, Ganten D, Lang RE, Suzuki F, Murakami K, Burbach JH, Ludwig G (1988) Quantification of renin mRNA in various mouse tissues by a novel solution hybridization assay. J Hypertens 6:247-52
Paul M, Wagner J, Dzau VJ (1993) Gene expression of the renin-angiotensin system in human tissues. Quantitative analysis by the polymerase chain reaction. J Clin Invest 91:2058-64
Paul M, Poyan Mehr A, Kreutz R (2006) Physiology of local renin-angiotensin systems. Physiol Rev 86:747-803
Peter S (1976) Ultrastructural studies on the secretory process in the epithelioid cells of the juxtaglomerular apparatus. Cell Tissue Res 168:45-53
Peti-Peterdi J (2006) Calcium wave of tubuloglomerular feedback. Am J Physiol Renal Physiol 291:F473-80
Peti-Peterdi J, Chambrey R, Bebok Z, Biemesderfer D, St John PL, Abrahamson DR, Warnock DG, Bell PD (2000) Macula densa Na^+/H^+ exchange activities mediated by apical NHE2 and basolateral NHE4 isoforms. Am J Physiol Renal Physiol 278:F452-63

Peti-Peterdi J, Komlosi P, Fuson AL, Guan Y, Schneider A, Qi Z, Redha R, Rosivall L, Breyer MD, Bell PD (2003) Luminal NaCl delivery regulates basolateral PGE_2 release from macula densa cells. J Clin Invest 112:76-82

Peti-Peterdi J, Fintha A, Fuson AL, Tousson A, Chow RH (2004) Real-time imaging of renin release in vitro. Am J Physiol Renal Physiol 287:F329-35

Pham I, Sediame S, Maistre G, Roudot-Thoraval F, Chabrier PE, Carayon A, Adnot S (1997) Renal and vascular effects of C-type and atrial natriuretic peptides in humans. Am J Physiol 273:R1457-64

Pieruzzi F, Munforti C, Di Blasio A, Busca G, Dadone V, Zanchetti A, Golin R (2002) Neuronal nitric oxide synthase and renin stimulation by sodium deprivation are dependent on the renal nerves. J Hypertens 20:2039-45

Piggott LA, Hassell KA, Berkova Z, Morris AP, Silberbach M, Rich TC (2006) Natriuretic peptides and nitric oxide stimulate cGMP synthesis in different cellular compartments. J Gen Physiol 128:3-14

Porter JP, Reid IA, Said SI, Ganong WF (1982) Stimulation of renin secretion by vasoactive intestinal peptide. Am J Physiol 243:F306-10

Porter JP, Said SI, Ganong WF (1983) Vasoactive intestinal peptide stimulates renin secretion in vitro: evidence for a direct action of the peptide on the renal juxtaglomerular cells. Neuroendocrinology 36:404-8

Prieto-Carrasquero MC, Harrison-Bernard LM, Kobori H, Ozawa Y, Hering-Smith KS, Hamm LL, Navar LG (2004) Enhancement of collecting duct renin in angiotensin II-dependent hypertensive rats. Hypertension 44:223-9

Prieto-Carrasquero MC, Kobori H,Ozawa Y, Gutierrez A, Seth D, Navar LG (2005) AT1 receptor-mediated enhancement of collecting duct renin in angiotensin II-dependent hypertensive rats. Am J Physiol Renal Physiol 289:F632-7

Rakugi H, Nakamaru M, Saito H, Higaki J, Ogihara T (1988) Endothelin inhibits renin release from isolated rat glomeruli. Biochem Biophys Res Commun 155:1244-7

Rasch R, Jensen BL, Nyengaard JR, Skott O (1998) Quantitative changes in rat renin secretory granules after acute and chronic stimulation of the renin system. Cell Tissue Res 292:563-71

Re RN (2004) Tissue renin angiotensin systems. Med Clin North Am 88:19-38

Reid IA, Chou L (1995) Effect of blockade of nitric oxide synthesis on the renin secretory response to frusemide in conscious rabbits. Clin Sci (Lond) 88:657-63

Reinecke M, Forssmann WG (1988) Neuropeptide (neuropeptide Y, neurotensin, vasoactive intestinal polypeptide, substance P, calcitonin gene-related peptide, somatostatin) immunohistochemistry and ultrastructure of renal nerves. Histochemistry 89:1-9

Ren Y, Arima S, Carretero OA, Ito S (2002) Possible role of adenosine in macula densa control of glomerular hemodynamics. Kidney Int 61:169-76

Reudelhuber TL (2005) The renin-angiotensin system: peptides and enzymes beyond angiotensin II. Curr Opin Nephrol Hypertens 14:155-9

Richards AM, Tonolo G, Montorsi P, Finlayson J, Fraser R, Inglis G, Towrie A, Morton JJ (1988) Low dose infusions of 26- and 28-amino acid human atrial natriuretic peptides in normal man. J Clin Endocrinol Metab 66:465-72

Richter CM, Godes M, Wagner C, Maser-Gluth C, Herzfeld S, Dorn M, Priem F, Slowinski T, Bauer C, Schneider W, Neumayer HH, Kurtz A, Hocher B (2004) Chronic cyclooxygenase-2 inhibition does not alter blood pressure and kidney function in renovascular hypertensive rats. J Hypertens 22:191-8

Rodriguez-Puyol D, Arriba G, Blanchart A, Santos JC, Caramelo C, Fernandez-Cruz A, Hernando L, Lopez-Novoa JM (1986) Lack of a direct regulatory effect of atrial natriuretic factor on prostaglandins and renin release by isolated rat glomeruli. Biochem Biophys Res Commun 138:496-501

Rodriguez F, Llinas MT, Gonzalez JD, Rivera J, Salazar FJ (2000) Renal changes induced by a cyclooxygenase-2 inhibitor during normal and low sodium intake. Hypertension 36:276-81

Rohrwasser A, Morgan T, Dillon HF, Zhao L, Callaway CW, Hillas E, Zhang S, Cheng T, Inagami T, Ward K, Terreros DA, Lalouel JM (1999) Elements of a paracrine tubular renin-angiotensin system along the entire nephron. Hypertension 34:1265-74

Rohrwasser A, Ishigami T, Gociman B, Lantelme P, Morgan T, Cheng T, Hillas E, Zhang S, Ward K, Bloch-Faure M, Meneton P, Lalouel JM (2003) Renin and kallikrein in connecting tubule of mouse. Kidney Int 64:2155-62
Russ U, Rauch U, Quast U (1999) Pharmacological evidence for a KATP channel in renin-secreting cells from rat kidney. J Physiol 517 (Pt 3):781-90
Ryan GB, Alcorn D, Coghlan JP, Hill PA, Jacobs R (1982) Ultrastructural morphology of granule release from juxtaglomerular myoepithelioid and peripolar cells. Kidney Int Suppl 12:S3-8
Ryan MJ, Gross KW, Hajduczok G (2000) Calcium-dependent activation of phospholipase C by mechanical distension in renin-expressing As4.1 cells. Am J Physiol Endocrinol Metab 279:E823-9
Saez JC, Berthoud VM, Branes MC, Martinez AD, Beyer EC (2003) Plasma membrane channels formed by connexins: their regulation and functions. Physiol Rev 83:1359-400
Salazar FJ, Fiksen-Olsen MJ, Opgenorth TJ, Granger JP, Burnett JC, Jr., Romero JC (1986) Renal effects of ANP without changes in glomerular filtration rate and blood pressure. Am J Physiol 251:F532-6
Sällström J, Carlström M, Jensen BL, Skott O, Brown RD, Persson AE (2008) Neuronal nitric oxide synthase-deficient mice have impaired renin release but normal blood pressure. Am J Hypertens 21:111-6
Saussine C, Judes C, Massfelder T, Musso MJ, Simeoni U, Hannedouche T, Helwig JJ (1993) Stimulatory action of parathyroid hormone on renin secretion in vitro: a study using isolated rat kidney, isolated rabbit glomeruli and superfused dispersed rat juxtaglomerular cells. Clin Sci (Lond) 84:11-9
Schnermann J, Levine DZ (2003) Paracrine factors in tubuloglomerular feedback: adenosine, ATP, and nitric oxide. Annu Rev Physiol 65:501-29
Schnermann J, Weihprecht H, Briggs JP (1990) Inhibition of tubuloglomerular feedback during adenosine1 receptor blockade. Am J Physiol 258:F553-61
Scholz H, Kurtz A (1993) Involvement of endothelium-derived relaxing factor in the pressure control of renin secretion from isolated perfused kidney. J Clin Invest 91:1088-94
Scholz H, Kurtz A (1995) Differential regulation of cytosolic calcium between afferent arteriolar smooth muscle cells from mouse kidney. Pflugers Arch 431:46-51
Scholz H, Vogel U, Kurtz A (1993) In terrelation between baroreceptor and macula densa mechanisms in the control of renin secretion. J Physiol 469:511-24
Scholz H, Hamann M, Gotz KH, Kurtz A (1994) Role of calcium ions in the pressure control of renin secretion from the kidneys. Pflugers Arch 428:173-8
Schricker K, Kurtz A (1993) Liberators of NO exert a dual effect on renin secretion from isolated mouse renal juxtaglomerular cells. Am J Physiol 265:F180-6
Schricker K, Kurtz A (1996) Blockade of nitric oxide formation inhibits the stimulation of the renin system by a low salt intake. Pflugers Arch 432:187-91
Schricker K, Della Bruna R, Hamann M, Kurtz A (1994a) Endothelium derived relaxing factor is involved in the pressure control of renin gene expression in the kidney. Pflugers Arch 428:261-8
Schricker K, Hamann M, Kaissling B, Kurtz A (1994b) Renal autacoids are involved in the stimulation of renin gene expression by low perfusion pressure. Kidney Int 46:1330-6
Schricker K, Hamann M, Kaissling B, Kurtz A (1994c) Role of the macula densa in the control of renal renin gene expression in two-kidney/one-clip rats. Pflugers Arch 427:42-6
Schricker K, Potzl B, Hamann M, Kurtz A (1996) Coordinate changes of renin and brain-type nitric-oxide-synthase (b-NOS) mRNA levels in rat kidneys. Pflugers Arch 432:394-400
Schweda F, Kurtz A (2004) Cellular mechanism of renin release. Acta Physiol Scand 181:383-90
Schweda F, Riegger GA, Kurtz A, Kramer BK (2000) Store-operated calcium influx inhibits renin secretion. Am J Physiol Renal Physiol 279:F170-6
Schweda F, Wagner C, Kramer BK, Schnermann J, Kurtz A (2003) Preserved macula densa-dependent renin secretion in A1 adenosine receptor knockout mice. Am J Physiol Renal Physiol 284:F770-7
Schweda F, Kammerl M, Wagner C, Kramer BK, Kurtz A (2004a) Upregulation of macula densa cyclooxygenase-2 expression is not dependent on glomerular filtration. Am J Physiol Renal Physiol 287:F95-101

Schweda F, Klar J, Narumiya S, Nusing RM, Kurtz A (2004b) Stimulation of renin release by prostaglandin E_2 is mediated by EP2 and EP4 receptors in mouse kidneys. Am J Physiol Renal Physiol 287:F427-33

Schweda F, Segerer F, Castrop H, Schnermann J, Kurtz A (2005) Blood pressure-dependent inhibition of renin secretion requires A1 adenosine receptors. Hypertension 46:780-6

Schweda F, Friis U, Wagner C, Skott O, Kurtz A (2007) Renin release. Physiology (Bethesda) 22:310-9

Sharp MG, Fettes D, Brooker G, Clark AF, Peters J, Fleming S, Mullins JJ (1996) Targeted inactivation of the Ren-2 gene in mice. Hypertension 28:1126-31

Shesely EG, Maeda N, Kim HS, Desai KM, Krege JH, Laubach VE, Sherman PA, Sessa WC, Smithies O (1996) Elevated blood pressures in mice lacking endothelial nitric oxide synthase. Proc Natl Acad Sci USA 93:13176-81

Shi SJ, Nguyen HT, Sharma GD, Navar LG, Pandey KN (2001) Genetic disruption of atrial natriuretic peptide receptor-A alters renin and angiotensin II levels. Am J Physiol Renal Physiol 281:F665-73

Sigmund CD, Gross KW (1991) Structure, expression, and regulation of the murine renin genes. Hypertension 18:446-57

Silver MA (2006) The natriuretic peptide system: kidney and cardiovascular effects. Curr Opin Nephrol Hypertens 15:14-21

Singh I, Grams M, Wang WH, Yang T, Killen P, Smart A, Schnermann J, Briggs JP (1996) Coordinate regulation of renal expression of nitric oxide synthase, renin, and angiotensinogen mRNA by dietary salt. Am J Physiol 270:F1027-37

Siragy HM, Linden J (1996) Sodium intake markedly alters renal interstitial fluid adenosine. Hypertension 27:404-7

Skott O (1986) Episodic release of renin from single isolated superfused rat afferent arterioles. Pflugers Arch 407:41-5

Skott O, Briggs JP (1987) Direct demonstration of macula densa-mediated renin secretion. Science 237:1618-20

Smyth JT, Dehaven WI, Jones BF, Mercer JC, Trebak M, Vazquez G, Putney JW, Jr. (2006) Emerging perspectives in store-operated Ca^{2+} entry: roles of Orai, Stim and TRP. Biochim Biophys Acta 1763:1147-60

Söhl G, Willecke K (2004) Gap junctions and the connexin protein family. Cardiovasc Res 62:228-32

Stichtenoth DO, Marhauer V, Tsikas D, Gutzki FM, Frolich JC (2005) Effects of specific COX-2-inhibition on renin release and renal and systemic prostanoid synthesis in healthy volunteers. Kidney Int 68:2197-207

Struthers AD, Anderson JV, Payne N, Causon RC, Slater JD, Bloom SR (1986) The effect of atrial natriuretic peptide on plasma renin activity, plasma aldosterone, and urinary dopamine in man. Eur J Clin Pharmacol 31:223-6

Sun D, Samuelson LC, Yang T, Huang Y, Paliege A, Saunders T, Briggs J, Schnermann J (2001) Mediation of tubuloglomerular feedback by adenosine: evidence from mice lacking adenosine 1 receptors. Proc Natl Acad Sci USA 98:9983-8

Takagi M, Franco-Saenz R, Mulrow PJ (1988a) Effect of atrial natriuretic peptide on renin release in a superfusion system of kidney slices and dispersed juxtaglomerular cells. Endocrinology 122:1437-42

Takagi M, Matsuoka H, Atarashi K, Yagi S (1988b) Endothelin: a new inhibitor of renin release. Biochem Biophys Res Commun 157:1164-8

Takahashi N, Chernavvsky DR, Gomez RA, Igarashi P, Gitelman HJ, Smithies O (2000) Uncompensated polyuria in a mouse model of Bartter's syndrome. Proc Natl Acad Sci USA 97:5434-9

Takenaka T, Inoue T, Kanno Y, Okada H, Meaney KR, Hill CE, Suzuki H (2008) Expression and role of connexins in the rat renal vasculature. Kidney Int 73:415-22

Tan DY, Meng S, Manning RD, Jr. (1999) Role of neuronal nitric oxide synthase in Dahl salt-sensitive hypertension. Hypertension 33:456-61

Taugner R, Hackenthal E (1988) On the character of the secretory granules in juxtaglomerular epithelioid cells. Int Rev Cytol 110:93-131

Taugner R, Schiller A, Kaissling B, Kriz W (1978) Gap junctional coupling between the JGA and the glomerular tuft. Cell Tissue Res 186:279-85
Taugner R, Hackenthal E, Inagami T, Nobiling R, Poulsen K (1982) Vascular and tubular renin in the kidneys of mice. Histochemistry 75:473-84
Taugner R, Buhrle CP, Nobiling R(1984a) Ultrastructural changes associated with renin secretion from the juxtaglomerular apparatus of mice. Cell Tissue Res 237:459-72
Taugner R, Kirchheim H, Forssmann WG (1984b) Myoendothelial contacts in glomerular arterioles and in renal interlobular arteries of rat, mouse and Tupaia belangeri. Cell Tissue Res 235:319-25
Taugner R, Nobiling R, Metz R, Taugner F, Buhrle C, Hackenthal E (1988) Hypothetical interpretation of the calcium paradox in renin secretion. Cell Tissue Res 252:687-90
Tharaux PL, Dussaule JC, Pauti MD, Vassitch Y, Ardaillou R, Chatziantoniou C (1997) Activation of renin synthesis is dependent on intact nitric oxide production. Kidney Int 51:1780-7
Thomson S, Bao D, Deng A, Vallon V (2000) Adenosine formed by 5'-nucleotidase mediates tubuloglomerular feedback. J Clin Invest 106:289-98
Thrasher TH (1994) Baroreceptor regulation of vasopressin and renin secretion: low-pressure versus high-pressure receptors. Front Neuroendocrinol 15:157-96
Thurau K, Schnermann J (1965) [The sodium concentration in the macula densa cells as a regulating factor for glomerular filtration (micropuncture experiments).]. Klin Wochenschr 43:410-3
Tigerstedt R, Bergmann P (1898) Niere und Kreislauf. Vol 8. 223-271
Tipnis SR, Hooper NM, Hyde R, Karran E, Christie G, Turner AJ (2000) A human homolog of angiotensin-converting enzyme. Cloning and functional expression as a captopril-insensitive carboxypeptidase. J Biol Chem 275:33238-43
Tkacs NC, Kim M, Denzon M, Hargrave B, Ganong WF (1990) Pharmacological evidence for involvement of the sympathetic nervous system in the increase in renin secretion produced by a low sodium diet in rats. Life Sci 47:2317-22
Toffelmire EB, Slater K, Corvol P, Menard J, Schambelan M (1989) Response of plasma prorenin and active renin to chronic and acute alterations of renin secretion in normal humans. Studies using a direct immunoradiometric assay. J Clin Invest 83:679-87
Tojo A, Madsen KM, Wilcox CS (1995) Expression of immunoreactive nitric oxide synthase isoforms in rat kidney. Effects of dietary salt and losartan. Jpn Heart J 36:389-98
Tomura Y, Hisa H, Satoh S (1990) Effects of atrial natriuretic peptide on phenylephrine-induced renin release in dogs. Clin Exp Pharmacol Physiol 17:381-4
Ujiie K, Yuen J, Hogarth L, Danziger R, Star RA (1994) Localization and regulation of endothelial NO synthase mRNA expression in rat kidney. Am J Physiol 267:F296-302
Vander AJ (1968) Direct effects of prostaglandin on renal function and renin release in anesthetized dog. Am J Physiol 214:218-21
Vander AJ, Geelhoed GW (1965) Inhibition of renin secretion by angiotensin. II. Proc Soc Exp Biol Med1 20:399-403
Vandongen R, Peart WS, Boyd GW(1973) Adrenergic stimulation of renin secretion in the isolated perfused rat kidney. Circ Res 32:290-6
Veerappan A, Reid AC, Estephan R, O'Connor N, Thadani-Mulero M, Salazar-Rodriguez M, Levi R, Silver RB (2008) Mast cell renin and a local renin-angiotensin system in the airway: Role in bronchoconstriction. Proc Natl Acad Sci USA 105:1315-20
Vikse A, Kiil F (1985) Enhancement of renal prostaglandin E_2 and renin release by autoregulatory dilation of preglomerular vessels in dogs. Ren Physiol 8:169-78
Vikse A, Bugge J, Dahl E, Kiil F (1985) Dissociation between renal prostaglandin E_2 and renin release. Effects of glucagon, dopamine and cyclic AMP in dogs. Acta Physiol Scand 125:619-26
Villarreal D, Freeman RH, Davis JO, Verburg KM, Vari RC (1986) Renal mechanisms for suppression of renin secretion by atrial natriuretic factor. Hypertension 8:II28-35
Wagner C (2008) Function of connexins in the renal circulation. Kidney Int 73:547-55
Wagner C, Kurtz A (1998) Regulation of renal renin release. Curr Opin Nephrol Hypertens 7:437-41

Wagner C, Pfeifer A, Ruth P, Hofmann F, Kurtz A (1998) Role of cGMP-kinase II in the control of renin secretion and renin expression. J Clin Invest 102:1576-82
Wagner C, Hinder M, Kramer BK, Kurtz A (1999) Role of renal nerves in the stimulation of the renin system by reduced renal arterial pressure. Hypertension 34:1101-5
Wagner C, Godecke A, Ford M, Schnermann J, Schrader J, Kurtz A (2000) Regulation of renin gene expression in kidneys of eNOS- and nNOS-deficient mice. Pflugers Arch 439:567-72
Wagner C, de Wit C, Kurtz L, Grunberger C, Kurtz A, Schweda F (2007) Connexin40 is essential for the pressure control of renin synthesis and secretion. Circ Res 100:556-63
Wang JL, Cheng HF, Harris RC (1999) Cyclooxygenase-2 inhibition decreases renin content and lowers blood pressure in a model of renovascular hypertension. Hypertension 34:96-101
Wang PH, Do YS, Macaulay L, Shinagawa T, Anderson PW, Baxter JD, Hsueh WA (1991) Identification of renal cathepsin B as a human prorenin-processing enzyme. J Biol Chem 266:12633-8
Wang T, Brown MJ (2004) Differential expression of adenylyl cyclase subtypes in human cardiovascular system. Mol Cell Endocrinol 223:55-62
Watkins BE, Davis JO, Lohmeier TE, Freeman RH (1976) Intrarenal site of action of calcium on renin secretion in dogs. Circ Res 39:847-53
Weber T, Zemelman BV, McNew JA, Westermann B, Gmachl M, Parlati F, Sollner TH, Rothman JE (1998) SNAREpins: minimal machinery for membrane fusion. Cell 92:759-72
Weihprecht H, Lorenz JN, Schnermann J, Skott O, Briggs JP (1990) Effect of adenosine1-receptor blockade on renin release from rabbit isolated perfused juxtaglomerular apparatus. J Clin Invest 85:1622-8
Weinberger MH, Aoi W, Henry DP (1975) Direct effect of beta-adrenergic stimulation on renin release by the rat kidney slice in vitro. Circ Res 37:318-24
Wilcox CS, Welch WJ, Murad F, Gross SS, Taylor G, Levi R, Schmidt HH (1992) Nitric oxide synthase in macula densa regulates glomerular capillary pressure. Proc Natl Acad Sci USA 89:11993-7
Wilcox CS, Welch WJ, Schreiner GF, Belardinelli L (1999) Natriuretic and diuretic actions of a highly selective adenosine A1 receptor antagonist. J Am Soc Nephrol 10:714-20
Yang T, Endo Y, Huang YG, Smart A, Briggs JP, Schnermann J (2000a) Renin expression in COX-2-knockout mice on normal or low-salt diets. Am J Physiol Renal Physiol 279:F819-25
Yang T, Park JM, Arend L, Huang Y, Topaloglu R, Pasumarthy A, Praetorius H, Spring K, Briggs JP, Schnermann J (2000b) Low chloride stimulation of prostaglandin E_2 release and cyclooxygenase-2 expression in a mouse macula densa cell line. J Biol Chem 275:37922-9
Yao J, Suwa M, Li B, Kawamura K, Morioka T, Oite T (2003) ATP-dependent mechanism for coordination of intercellular Ca^{2+} signaling and renin secretion in rat juxtaglomerular cells. Circ Res 93:338-45
Yokosawa N, Takahashi N, Inagami T, Page DL (1979) Isolation of completely inactive plasma prorenin and its activation by kallikreins. A possible new link between renin and kallikrein. Biochim Biophys Acta 569:211-9
Zhang J, Hill CE (2005) Differential connexin expression in preglomerular and postglomerular vasculature: accentuation during diabetes. Kidney Int 68:1171-85
Zhang Y, Morgan T (1994) Role of the macula densa in renin synthesis and secretion. Am J Hypertens 7:448-52
Zou AP, Wu F, Li PL, Cowley AW, Jr. (1999) Effect of chronic salt loading on adenosine metabolism and receptor expression in renal cortex and medulla in rats. Hypertension 33:511-6

Shaping Membranes for Endocytosis

M. Krauss and V. Haucke

Abstract Endocytosis is essential for virtually all eukaryotic cells to internalize nutrients, antigens, pathogens, and cell surface receptors from the plasma membrane into membrane-bounded, endocytic vesicles to regulate cell homeostasis, cell signaling, and development. Distinct mechanisms mediate endocytic uptake of a large variety of distinctly sized cargoes ranging from small molecules to viruses or bacteria. Common to all of these endocytic pathways is the deformation of the plasma membrane by intracellular factors including scaffolding proteins, amphipathic peripheral membrane proteins, and lipid-modifying enzymes. In this review we summarize how different cargoes exploit distinct pathways for cell entry, and how proteins assist the generation of curved membrane domains during internalization.

Abbreviations ALPS: ArfGAP1 lipid-packing sensor; Arf: ADP-ribosylation factor; BAR: Bin-amphiphysin-Rvs; CLASP: Clathrin-associated sorting protein; CME: Clathrin-mediated endocytosis; ENTH: Epsin N-terminal homology; FCH: Fes/CIP4 homology; GAP: GTPase-activating protein; GPI: Glycosyl phosphatidylinositol; PI: Phosphoinositide; PI(4,5)P_2: Phosphatidylinositol 4,5-bisphosphate; PIPK I: Phosphatidylinositol 4-phosphate 5-kinase; PLD: Phospholipase D; PX: Phox: homology; SNX9: Sorting nexin 9; STx: Shiga toxin; SV40: Simian virus 40; TIRF: Total internal reflection fluorescence

Virtually all eukaryotic cells use endocytic mechanisms to retrieve nutrients, antigens, pathogens, and cell surface receptors from the plasma membrane. In all of these cases transport is mediated by membrane-bounded tubular or vesicular carriers that are formed at distinct sites at the plasmalemma and are then trafficked

M. Krauss (✉)
Institute of Chemistry and Biochemistry, Department of Membrane Biochemistry
Freie Universität Berlin
Takustraße 6, 14195
Berlin, Germany
e-mail: mkrauss@chemie.fu-berlin.de

Rev Physiol Biochem Pharmacol, doi: 10.1007/112_2008_2

along intracellular endosomal pathways. Tubulovesicular membrane structures are also frequently observed during endosomal progression and in endosomal recycling. Multiple distinct pathways for cargo uptake and intracellular targeting exist that differ in their dependence on coat proteins implicated in budding and on GTPases frequently regulating lipid metabolism and actin remodeling. In the case of signaling receptors, carrier formation might also be modulated by components of activated intracellular signaling cascades.

One hallmark of endocytosis is the generation of invaginated membrane areas with positive or negative curvature around forming buds or tubules and at the neck connecting budding vesicles to the plasma membrane (McMahon and Gallop 2005). The degree of curvature varies significantly at different stages of vesicle formation, and is most probably regulated by a stepwise entry of curvature inducing or stabilizing proteins into the forming bud or tubule. Multiple mechanisms have been described that support bending of membranes. These include scaffolding proteins exemplified by the polymerization of coat proteins including clathrin, insertion of hydrophobic moieties into one leaflet of the lipid bilayer, i.e., the hydrophobic face of amphipathic helices, and changes in lipid composition. Moreover, mechanical forces exerted by the actin cytoskeleton and by motor proteins also support the generation of curved membrane domains.

Cargoes can significantly differ in size, ranging from small molecules like receptor ligands up to large particles, including viruses, bacteria, and extracellular aggregates. Accordingly, binding of ligands to the cell surface will have to activate appropriate intracellular machineries that can then trigger changes in membrane shape to facilitate their uptake. In this review we summarize how different cargoes exploit distinct pathways for cell entry (Fig. 1), and how proteins assist the generation of curved membrane domains during internalization.

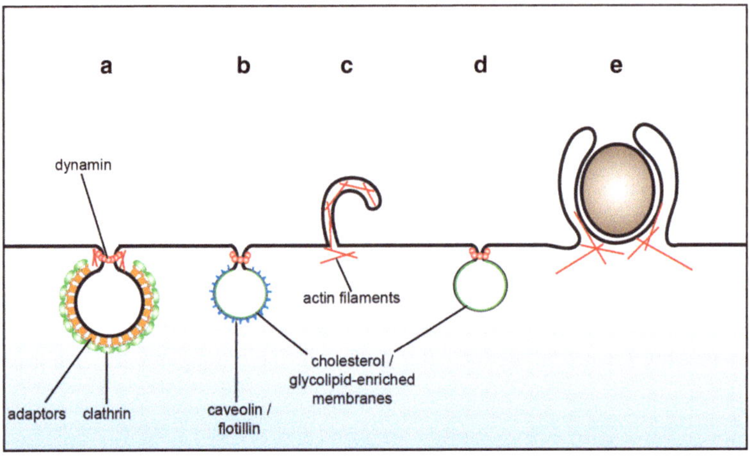

Fig. 1 a–e Multiple pathways for endocytic uptake of cargo proteins into the mammalian cell. They differ with regard to the class of cargo (receptors, ligands, lipids, and lipid-associated molecules), the mechanism of vesicle formation (coats, dependence on small GTPases and dynamin), and the size of the endocytic vesicle finally pinched off into the cytoplasm. **a** Clathrin-mediated

Taking a Trip into the Cell

Clathrin-Mediated Endocytosis

Clathrin-mediated endocytosis (CME) is a universal process essential for the uptake of nutrients such as the iron carrier transferrin or low-density lipoprotein particles that bind their respective receptors. Furthermore, CME modulates cell–cell communication events by controlling the levels of surface-exposed membrane proteins and retrieval of activated signaling receptors from the plasma membrane. Thereby it contributes to cellular differentiation processes for tissue development. In addition, clathrin is required for the internalization of numerous pumps and transporters for ions and small metabolites involved in cell homeostasis, and synaptic transmission in neurons. Moreover, CME is required for the efficient recycling of presynaptic vesicles during sustained neuronal activity as well as to regulate the levels of surface-active neurotransmitter receptors on the postsynaptic side. Finally, some toxins and viruses undergo receptor-mediated internalization via CME.

CME involves the assembly of cytosolic coat and accessory proteins at phosphatidylinositol 4,5-bisphosphate (PI(4,5)P_2)-enriched plasmalemmal sites (Fig. 1a). Coat polymerization serves as a mechanical means to stabilize the deformed plasma membrane and to concentrate cargo membrane proteins within the nascent buds. Clathrin-coated buds eventually pinch off from the plasma membrane in a reaction catalyzed by the GTPase activity of dynamin. The released vesicle is rapidly uncoated to allow its fusion with endosomal membranes. Vesicles generated by CME are strikingly uniform in size (about 80–100 nm in diameter), indicating that tight regulatory mechanisms must govern membrane bending and vesicle formation. Accordingly, curvature acquisition is the result of a complex interplay between numerous protein and lipid factors (see below).

Clathrin itself is an oligomer consisting of three heavy and three light chains which assemble into three-legged triskelia. Clathrin recruitment to the plasma membrane is mediated by adaptor proteins, including the heterotetrameric complex AP-2 and numerous clathrin-associated sorting proteins (CLASPs). AP-2 serves as a major protein–protein and protein–lipid interaction 'hub' providing multiple binding sites for PI(4,5)P_2, cargo proteins, clathrin, and accessory proteins (Edeling et al. 2006). Cargo recognition depends on short peptide motifs exposed in the

Fig. 1 a–e (continued) endocytosis employs adaptor proteins which link cargo proteins to the clathrin scaffold covering the forming endocytic vesicle. **b** Caveolae are formed by clustered caveolin proteins, and are enriched in cholesterol, sphingolipids and glycolipids presumably concentrated in membrane microdomains. Flotillins might serve similar functions. **c** During macropinocytosis actin-driven membrane protrusions are formed that non-selectively enclose extracellular fluid when they collapse back onto, and fuse with, the plasma membrane. **d** Some cargoes associate with membrane microdomains (*green*) that are internalized in a clathrin- and caveolin-independent manner. Different modes of internalization can be distinguished with regard to their dependence on dynamin. **e** Phagocytosis mediates internalization by membrane protrusions that are formed around a large, ligand-coated particle

cytoplasmic tails of transmembrane proteins. These include tyrosine-based motifs (YXXF, with Ø representing a large hydrophobic amino acid), acidic-cluster dileucine motifs (D/EXXXLL/I), and stretches of basic amino acids (Jung et al. 2007). These motifs may target membrane cargo for constitutive CME, as in the case of the transferrin receptor, or function upon activation of signaling receptors such as the epidermal growth factor receptor.

Besides AP-2 and clathrin, numerous accessory factors have been identified that aid the formation and maturation of the newly forming vesicles. These accessory factors are recruited at distinct stages of clathrin-coated vesicle formation and assist in the selection and concentration of cargo proteins, curvature acquisition, lipid metabolism, or serve as scaffolding factors. Accessory proteins may also serve as (usually monomeric) adaptor proteins (CLASPs) that drive efficient sorting of specific cargoes into clathrin-coated pits (Bonifacino and Traub 2003). Like AP-2, CLASPs recognize short peptide motifs within select cargo that function as endocytosis signals and in some cases are found in combination with conventional endocytic motifs. For instance, Numb, ARH, and Dab2 have been implicated in CME of NPXY-containing proteins, i.e., present in the cytoplasmic tails of integrins, APP, or of the receptors for low-density lipoprotein and some growth factors. Stonin 2 was identified as a CLASP for the internalization of presynaptic vesicles via association with synaptotagmin (Grass et al. 2004; Jung et al. 2007), and arrestins play a role in the endocytosis of G-protein coupled receptors (Wolfe and Trejo 2007). Importantly, at least some cargo proteins appear to concentrate in discrete clathrin-coated pits and display different kinetics of internalization (Keyel et al. 2006; Lakadamyali et al. 2006; Puthenveedu and von Zastrow 2006), suggesting specializations of the CME pathway. For example, ß-arrestin-mediated internalization of G-protein coupled receptors is delayed when compared with the rate of constitutively endocytosed transferrin receptors. Moreover, endosomal progression differs between constitutively endocytosed cargo and signaling receptors. The prolonged surface retention of G-protein coupled receptors has been attributed to receptor association with scaffolding proteins. Other activated signaling receptors might display similar properties, dependent on the individual signal transduction cascades they elicit.

Since several proteins involved in CME, including AP-2, AP180/CALM, epsin, numb, Dab2, etc., harbor binding domains for $PI(4,5)P_2$ (Haucke 2005), efficient coat formation and release depend on the coordinated activities of phosphoinositide (PI) kinases and phosphatases (Krauss and Haucke 2007). Moreover, CME is accompanied by local polymerization of actin, the precise role of which remains somewhat uncertain. These mechanisms will be described in more detail below.

Strikingly, some invasive bacteria rely on CME to enter cells, although their size appears to exceed the shape restrictions imposed by conventional clathrin-coated vesicles. *Listeria* bind to the cell surface receptor tyrosine kinase Met that undergoes activation-dependent ubiquitination thereby facilitating clathrin-dependent internalization of the invading bacteria (Bonazzi and Cossart 2006; Veiga and Cossart 2005). Thus, *Listeria* must trigger specific adaptations of the clathrin machinery to accommodate the bacterial cell. It will be interesting to see whether formation of coated pits at sites of bacterial entry requires specialized clathrin adaptors, the recruitment of which might be initiated by activation of the Met-signaling cascade. In this context it seems worth noting that *Listeria* have been shown to recruit E-cadherins and downstream components such as a- and ß-catenins and myosin VIIa to sites of cell entry (Lecuit et al. 2000). Therefore, at least in epithelial cells, E-cadherin-initiated changes of the actin cytoskeleton and motor-applied forces support clathrin-dependent listerial cell entry.

Caveolae-Mediated Endocytosis

Caveolae are highly abundant in endothelial and smooth muscle cells, adipocytes, and fibroblasts where they appear as flask-shaped invaginations of the plasma membrane of approximately 60–80 nm in diameter (Fig. 1b). In contrast to clathrin-coated structures they lack a characteristic coat. However, they contain high levels of caveolins, the ablation of which causes loss of morphologically recognizable caveolae from these cell types. Caveolins display an unusual topology with their N- and C-termini oriented towards the cytoplasm and a long intramembrane domain which presumably folds into a hairpin-loop-like structure protruding into the inner leaflet of the lipid bilayer (Cohen et al. 2004). Caveolins are post-translationally modified by palmitoylation, which enhances their anchoring to cellular membranes and may aid their partitioning into cholesterol- and glycosphingolipid-enriched plasma membrane microdomains.

If, and how, caveolae participate in endocytic events is a matter of debate. Endothelial cells lacking caveolin cannot bind to and internalize serum albumin (Razani et al. 2002). Yet, caveolin-deficient mice display normal serum and interstitial levels of albumin (Drab et al. 2001), indicating that alternative routes for albumin transcytosis must exist. Caveolae have also been implicated in endocytosis of simian virus 40 (SV40), which initially attaches to the ganglioside GM1 at the cell surface (Pelkmans et al. 2001). Virus binding activates intracellular signaling cascades that involve tyrosine kinases associated with cholesterol/glycosphingolipid microdomains and PI kinases (Pelkmans and Helenius 2002). Internalization of SV40 is dependent on dynamin as evidenced by overexpression of a GTPase-defective dynamin mutant resulting in the accumulation of ligand-engaged caveolae that remain attached to the plasma membrane through an elongated neck (Hinshaw 2000). Caveolin-dependent endocytosis is associated with remodeling of the cortical actin cytoskeleton. Similar to SV40, caveolae have been implicated in cellular uptake of GM1-bound cholera toxin (Lencer et al. 1999). Surprisingly, endocytosis of SV40 (Damm et al. 2005) and cholera toxin (Kirkham et al. 2005) persists in caveolin-deficient cells suggesting that alternative non-caveolar uptake routes must exist.

Additional functions for caveolins have also been proposed. Based on the static appearance of caveolae (Thomsen et al. 2002), and the finding that overexpression of caveolin inhibits caveolar uptake (Le et al. 2002), it was suggested that caveolae might serve as stores for sphingolipids and cholesterol. According to this hypothetical model caveolae would ensure supply of lipids for caveolin- and clathrin-independent endocytosis events. Alternatively, caveolins might be utilized to internalize these lipids from the plasma membrane and replenish lipid pools within intracellular membranes. A direct involvement of caveolin in internalization processes remains to be proven.

Clathrin- and Caveolin-Independent Endocytosis

Some proteins or small molecules enter cells by clathrin- and caveolin-independent pathways (Fig. 1c, d). These routes have been classified according to the requirements for dynamin during fission, and according to the regulatory roles of small GTPases (RhoA, Cdc42, Arf) (Mayor and Pagano 2007). Strikingly, none of these pathways appears to depend on adaptor or coat proteins. Internalization therefore must involve distinct mechanisms of cargo concentration and curvature generation. One possibility is that cargo associates with membrane microdomains enriched in cholesterol and glycosphingolipids. These microdomains are relatively small (40–50 nm in diameter), highly mobile, and rapidly exchange individual constituents with the surrounding membrane (Edidin 2001; Simons and Ikonen 1997). Assembly of such microdomains might then lead to the formation of an endocytic vesicle. Microdomains (also sometimes referred to as 'detergent-resistant membranes' based on the finding that their protein components remain partially insoluble in non-ionic detergents at 4°C) have been postulated to constitute sorting and signaling platforms via cholesterol- and glycosphingolipid-dependent trapping of lipid-associated membrane proteins (Harder 2004).

One example of cargo internalized via a clathrin- and caveolin-independent route is interleukin-2 (Lamaze et al. 2001) that associates with detergent-resistant membranes upon ligand binding from where it is endocytosed. Internalization is unaffected by inhibitors of CME (i.e., clathrin/AP2 sequestration via overexpression of dominant-negative mutants of Eps15), yet depends on dynamin function (Fig. 1d). In addition, endocytosis of interleukin-2 receptors is abolished by overexpression of constitutively active mutants of RhoA and Rac. Both GTPases are involved in remodeling of the actin cytoskeleton and might thereby coordinate plasma membrane invagination. Other actin-regulatory GTPases such as Cdc42 have also been shown to affect clathrin- and caveolin-independent uptake events. Cdc42 regulates endocytosis of bacterial toxins, including cholera toxin B, and of glycosyl phosphatidylinositol (GPI)-anchored proteins by induction of vacuoles (Kirkham et al. 2005; Sabharanjak et al. 2002). These contain large volumes of fluid phase, thereby constituting one form of macropinocytic uptake (Fig. 1c).

Recent studies implicate members of the Arf family of small GTPases, especially Arf1, in the uptake of GPI-anchored proteins and fluid phase markers (Kumari and Mayor 2008). Arf1-depletion inhibits internalization of CFP-GPI and fluorescent dextran, but does not affect other clathrin-dependent or -independent pathways. Arf1 recruits the RhoGAP ARHGAP10 to the plasmalemma, thereby stimulating Cdc42 recruitment and activation at sites of endocytosis triggering actin rearrangements. Similarly, Arf6 has been implicated in clathrin-independent and cholesterol-dependent endocytosis of MHC class I (Naslavsky et al. 2004), MHC class II (Walseng et al. 2008), and of M2-muscarinic acetylcholine receptors (Delaney et al. 2002). Given the extensive structural homology between Arf1 and Arf6 in their active conformations (Pasqualato et al. 2001), one might speculate that both GTPases share overlapping functions during clathrin-independent internalization of distinct cargoes. Indeed, activation of Arf6 in overexpressing cells leads to the formation of actin-dependent protrusions and stimulates macropinocytosis (Fig. 1c) (Radhakrishna et al. 1996). In general, Arf proteins fulfill complex functions, including the modulation of the actin cytoskeleton (predominantly through the regulation of Rho GTPases; Myers and Casanova 2008), and the stimulation of lipid-modifying enzymes (phospholipases and PI kinases; see below), and have therefore been implicated in several actin-dependent endocytic events (Donaldson 2003).

More recently, flotillin (reggie) proteins have been proposed to function in dynamin- and caveolin-independent endocytosis (Glebov et al. 2006). Flotillin-1 was identified in a screen for proteins associated with endosomes formed during fluid uptake of paramagnetic dextran. Depletion of flotillin-1 significantly impairs uptake of cholera toxin and GPI-anchored proteins. Flotillin-1 appears to function in concert with the related membrane protein flotillin-2. It has been suggested that both proteins co-assemble, thereby forming dynamic membrane microdomains of high curvature that bud into the cell (a mechanism similar to that depicted in Figs. 1b and 2C). Flotillin microdomains resemble caveolae in their ultrastructural organization, yet are functionally distinct (Frick et al. 2007). Flotillins share the topological features of caveolins with two putative hairpin-like loops protruding into one leaflet of the lipid bilayer. As they also partition into detergent-resistant membranes, it is assumed that flotillins represent structural components aiding the formation of highly curved membranes during distinct endocytic events. Whether flotillins form part of a larger, more complex endocytic machine remains to be determined.

Phagocytosis

Phagocytosis is a process that serves to dispose pathogens like bacteria or yeast, or large debris left over from apoptotic cells. Only some mammalian cells are specialized in phagocytic uptake, including monocytes, neutrophils, and macrophages. Phagocytosis is initiated when cell-surface receptors engage in interactions with their specific ligands. These ligands can be innate components of a pathogen's surface (termed opsonins) or pathogen-associated molecules contributed by the host. Correspondingly, a plethora of receptors exist that recognize sugar residues, phosphatidylserine, elements of the complement system, or immunoglobulins. Upon binding to their ligands, cell-surface receptors activate intracellular signaling cascades that frequently lead to the activation of PI kinases and phospholipases, members of the Rho and Arf families of GTPases, and of protein kinases including protein kinase C, and the tyrosine kinases Src and Syk (Chimini and Chavrier 2000; May and Machesky 2001; Niedergang and Chavrier 2004). Frequently, several receptors engage in particle binding simultaneously, provoking complex intracellular responses. Phagocytosis is accompanied by extensive remodeling of the plasmalemma to shape the phagocytic cup. This includes plasma membrane fusion events with intracellular membranes such as endosomes/lysosomes, secretory organelles, and perhaps the endoplasmic reticulum (Huynh et al. 2007), as well as local changes in phospholipid composition. RhoA and Cdc42 activities are modulated during engulfment of opsonized particles and induce the formation of actin filaments around the phagocytic cup. In addition, myosins appear to assist membrane deformation around its target and in membrane closure (Vicente-Manzanares and Sanchez-Madrid 2004). Interestingly, expression of mutant dynamin impairs the generation of membrane protrusions around the opsonized particle (Gold et al. 1999). This observation most probably reflects an involvement of dynamin during recruitment of actin regulators. Indeed, dynamin is found in actin-based structures during early stages of phagocytosis, together with other actin-associated proteins such as cortactin, Abp1, profilin, and syndapin (Orth and McNiven 2003).

Generating Membrane Curvature at the Plasma Membrane

All of the endocytic mechanisms described above require the deformation of the plasma membrane into tubular or bud-like structures of distinct curvatures (McMahon and Gallop 2005). Membrane bending involves the activities of curvature-inducing or stabilizing proteins as well as modification or clustering of membrane lipids. In the following section we outline some of the basic mechanisms (Fig. 2) contributing to the formation of curved membrane domains and use this information to derive some general principles.

Asymmetric Membrane Expansion: Dipping Fingers into the Sea

As outlined above, during vesicle or tubule formation cytoplasmic coat proteins are recruited to the plasma membrane at sites of cargo binding. At early stages these membrane areas will be flat. Endocytic proteins may initially disturb membrane structure locally by interacting with the charged surface of the lipid bilayer and concomitant insertion of hydrophobic moieties into the cytoplasmic leaflet of the membrane (Fig. 2a). These moieties usually consist of one or a few bulky hydrophobic amino acid side chains that are in some cases accompanied by lipid anchors attached in close proximity. Whereas the association of single proteins with the membrane is expected to create only subtle effects, curvature induction might

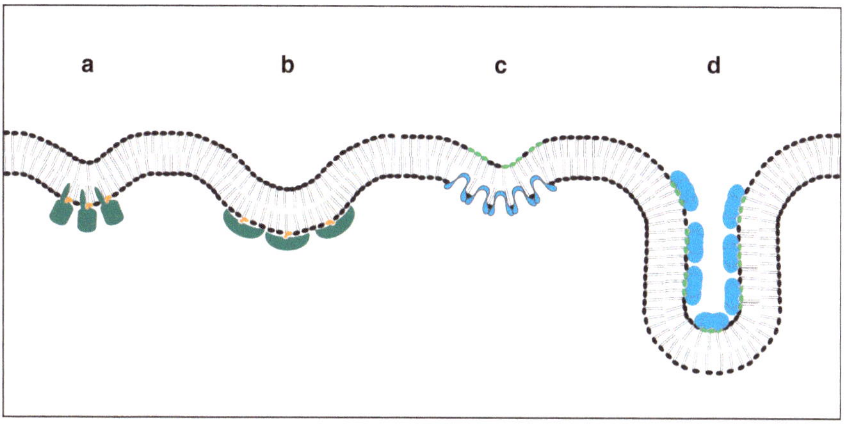

Fig. 2 a–d Different modes to create and/or sense membrane curvature. Initial recruitment of proteins to the plasma membrane is frequently assisted by negatively charged lipid headgroups, in particular phosphatidylinositol 4,5-bisphosphate (*orange*). **a** The insertion of hydrophobic moieties into the lipid bilayer (for instance amphipathic helices) generates an asymmetric expansion of the inner leaflet. **b** BAR domains (*green*) form banana-shaped dimers and can sense membrane domains of appropriate curvature. Subgroups of BARs can also generate membrane curvature. **c** The hydrophobic domains of some membrane proteins, such as caveolins and flotillins, form hairpin-loop-like intramembrane structures (*blue*) that penetrate into the inner leaflet of the lipid bilayer. By clustering together with associated lipid molecules (in particular cholesterol and glycolipids) they may generate flask-shaped membrane invaginations. **d** Some cargoes induce plasma membrane invaginations without the apparent aid of cytoplasmic coat proteins. The B-subunit of Shiga toxin (*blue*) forms pentamers, each of which can mediate binding of several molecules of globotriaosyl ceramide (*green*). Thus, oligomerization is accompanied by clustering of lipids and thereby favors the generation of negative membrane curvature

be potentiated by protein oligomerization and by combination with other membrane bending activities such as the stabilization of bent membranes by peripheral scaffolding proteins or asymmetric clustering of lipid molecules.

A well-studied example for this class of membrane tubulating proteins is provided by epsins and their homologues. Epsins are monomeric adaptors implicated in sorting of ubiquitinated cargo at the plasma membrane (Bonifacino and Traub 2003; Hicke and Dunn 2003). The carboxy-terminal domains of epsins 1–3 are largely unstructured, but include ubiquitin interacting motifs, NPF-motifs that mediate interactions with eps15 homology domain containing proteins, and several motifs that drive association with the amino terminal ß-propeller domain of clathrin and with the α-ear domain of AP-2. The membrane-bending activity of epsin 1 is encoded within its amino-terminal ENTH domain (Table 1) which selectively interacts with $PI(4,5)P_2$, thereby targeting epsin to the plasma membrane (Ford et al. 2002). $PI(4,5)P_2$ binding triggers a conformational change within the ENTH domain that induces the formation of an additional amphipathic helix (helix 0) at its amino-terminal end. Helix 0 may then partially insert into the

Table 1 Properties of protein domains assisting vesicle formation at the plasma membrane

Domain	Properties
ENTH	Epsin N-terminal homology; confers $PI(4,5)P_2$ specificity during binding to the plasma membrane; $PI(4,5)P_2$ binding induces a conformational change triggering de novo formation of an N-terminal amphipathic α-helix which partially inserts into the membrane
NBAR	Homology to Bin, amphiphysin, Rvs; helical coiled-coil domain forming a banana-shaped dimer that associates with lipids through its concave surface; thought to sense or induce positive membrane curvature
FBAR	Extended FCH (Fes/CIP4 homology) domain that assembles into a BAR domain; induces the formation of wider and more rigid tubules as compared to the NBAR domain
PX	Homology to Phox40; recognizes subsets of phosphoinositides and confers protein binding to distinct subcellular membrane surfaces
ALPS	ArfGAP1 lipid packing sensor; amphipathic α-helix inserted into one leaflet of the lipid bilayer; generated de novo upon protein binding to curved membranes

lipid bilayer inducing positive membrane curvature. In vitro, epsin 1 can generate membrane tubules or following addition of clathrin invaginated coated pits that arise from the concomitant recruitment and assembly of clathrin onto lipid monolayers. In vivo, epsin is found in AP-2-coated structures and this is dependent on the integrity of its $PI(4,5)P_2$-binding motif. In the absence of clathrin (Hinrichsen et al. 2006) the underlying plasma membrane remains flat, indicating that epsin alone at physiological concentrations is insufficient to create stably curved plasma membrane domains.

The question of how membranes are shaped during clathrin- or caveolin-independent events has remained elusive. As outlined above, members of the Arf family of small GTPases appear to be involved in many of these processes by recruiting effector proteins to the forming vesicle or tubule. Arf proteins contain an amino-terminal amphipathic helix that folds back onto the GTPase core in the inactive GDP-bound state. Upon GTP loading this helix has been proposed to detach from the GTPase core, thereby exposing hydrophobic residues and a myristate anchor to the membrane. The hydrophobic face of this helix is then inserted into the lipid bilayer (Antonny et al. 1997), assisted by electrostatic interactions with the membrane surface mediated by flanking basic residues. Membrane tubulating activities of Arf proteins, in particular of Arf1 at the Golgi, have recently been demonstrated in vitro and in vivo (Godi et al. 2004; Presley et al. 2002), and depend on the integrity of the amphipathic N-terminal helix (Krauss et al. 2008; Beck et al. 2008). A similar mechanism might underlie membrane deformation during Arf6-dependent macropinocytic uptake processes at the plasma membrane.

Stabilizing Curved Membrane Domains by Biological Basket Weaving

Initial observations based on electron microscopy had suggested that coat polymerization might represent a mechanism to stabilize curved membrane areas during formation of a clathrin-coated bud. Polygonal clathrin cages have been found to cover both flat and omega-shaped bud-like structures at the plasma membrane (Heuser 1980). Indeed, clathrin triskelia with the aid of adaptor proteins including epsins, AP-2, and CALM/ AP180 can assemble into closed polygonal cages that closely resemble clathrin-coated membrane buds seen by electron microscopy (Fotin et al. 2004a, 2004b; Musacchio et al. 1999). Accordingly, in clathrin-deficient fibroblasts plasma membrane areas coated with AP-2 and endocytic accessory proteins are visible but do not appear to be invaginated (Hinrichsen et al. 2006), indicating an essential role for clathrin during membrane deformation. Likewise, caveolae were first described as flask-shaped membrane invaginations (Palade and Bruns 1968) covered with spiral-like coats (Rothberg et al. 1992) comprised of caveolins. Unlike in the case of clathrin coats, no additional components have been identified. The number of flask-shaped caveolae closely correlates with the expression level of caveolins. Overexpression of caveolins leads to increased numbers of caveolae, whereas caveolae are absent from caveolin-depleted cells. Caveolins form stable oligomers of about 100 to 200 molecules within a single cholesterol-rich caveolar membrane domain (Monier et al. 1995; Pelkmans and Zerial 2005). Given the asymmetric topology of caveolins with protein domains inserting only into the cytoplasmic leaflet, oligomerization would generate positive curvature due to the local net increase introduced within the intracellular membrane surface (Fig. 2c). Flotillins display an analogous architecture and thus might act similarly (Frick et al. 2007).

Curving Through BARs

Several endocytic proteins including amphiphysin, endophilin, Toca/ FBP17/CIP4, syndapin/pacsin, and sorting nexins contain so-called BAR domains (Bin-amphiphysin-Rvs; Table 1) consisting of a triple helical coiled-coil that assembles into a banana-shaped dimer (Habermann 2004). Individual BAR domains associate with and stabilize membranes of distinct curvature (Fig. 2b) and therefore have been proposed to serve as curvature sensors that may act at late stages of vesicle formation. Alternatively, as BAR domains have been shown to confer membrane tubulating activities in vivo and in vitro, they might contribute to the invagination of the plasma membrane during endocytosis. Although BAR domain-containing proteins have been originally identified as parts of the machinery that forms clathrin-coated vesicles, they might serve mechanistically similar functions in clathrin-independent endocytosis, i.e., by associating with tubular membrane invaginations of appropriate diameter.

Amphiphysin was among the first endocytic proteins for which membrane tubulating activities have been demonstrated (Takei et al. 1999). In the case of amphiphysin the BAR domain is preceded by an additional amphipathic helix that may drive initial membrane deformation. The NBAR domain (comprising the actual BAR domain and the amino-terminal amphipathic helix) contains several basic patches exposed at the concave surface of the dimer and additional positively charged amino acid residues within flexible linkers between individual helices (Peter et al. 2004). By mutation of one or several of these residues, binding of amphiphysin to negatively charged liposomes is significantly impaired, suggesting that contacts of low specificity between basic amino acids and phospholipid head-groups are important for membrane association. Localization of amphiphysin at sites of endocytosis is mediated by direct interactions via conserved peptide motifs with clathrin and AP-2 (Slepnev et al. 2000) and by binding of its SH3 domain to proline-rich domain containing partner proteins including dynamin, endophilin, and synaptojanin (de Heuvel et al. 1997; Grabs et al. 1997). Interestingly, amphiphysin and dynamin can co-assemble into helical rings around tubular-shaped liposomes in vitro that closely resemble those observed in vivo (Takei et al. 1999). Furthermore, amphiphysin can potentiate the liposome-fragmenting activity of dynamin in vitro. Thus, it is tempting to speculate that amphiphysin couples membrane curvature induction and sensing to dynamin-mediated vesicle fission at the neck of clathrin-coated pits. A similar overall domain architecture comprising an N-terminal BAR- and a C-terminal SH3 domain is seen in the endocytic protein endophilin. The crystal structure of its BAR domain revealed a crescent shape, as in the case of amphiphysin, but also displayed important differences that might enhance its membrane-deforming properties (Masuda et al. 2006; Gallop et al. 2006), most notably a hydrophobic appendage on its concave surface. This hydrophobic ridge is thought to be inserted into the lipid bilayer, thereby enhancing endophilin-induced membrane deformation. Genetic as well as morphological studies in flies, worms, and lamprey suggest a tight functional link between endophilin and synaptojanin at synapses (Gad et al. 2000; Ringstad et al. 1999; Schuske et al. 2003; Verstreken et al. 2003). Collectively, these experiments suggest that endophilin serves to recruit the inositol phosphatase synaptojanin to endocytic sites of high local membrane curvature. Whether, and how, these activities are related to alteration of the actin cytoskeleton remains to be defined.

Mechanistic links between the actin cytoskeleton and BAR domain-mediated membrane deformation have also emerged for sorting nexins (SNX), a class of proteins comprising PI-recognizing Phox homology (PX; Table 1) as well as BAR and SH3 domains. For example, SNX9, a PI(4,5)P_2-associated clathrin/ AP-2 binding protein, via its SH3 domain forms complexes with dynamin and with N-WASP, a major regulator of ARP2/3-mediated actin nucleation (Lundmark and Carlsson 2002, 2003, 2004; Pylypenko et al. 2007). SNX9 overexpression in living cells triggers the formation of tubular structures and this activity is dependent on the integrity of its BAR and PX domains as well as on its Arp2/3 binding linker region (Shin et al. 2008). The membrane deforming activity of SNX9 is further supported by an amphipathic helix lying at the distal ends of each PX domain that might become inserted into the plasmalemma upon BAR domain binding to acidic phospholipids.

Structural predictions have unraveled FCH (Fes/CIP4 homology) proteins as an additional subfamily of BAR domain-containing proteins (Table 1). Crystallographic (Shimada et al. 2007) and cell biological data (Itoh and De Camilli 2006) show that the FCH domain together with another predicted coiled-coil region assembles into a single unit structurally and functionally resembling BAR domains, hence referred to as FBAR domain. FBAR-containing proteins appear to associate with much wider tubular membranes of about 600 Å in diameter. Cryo-electron microscopy studies have revealed the structural basis for the distinctive FBAR protein-mediated membrane tubulation properties. The initial contact with the membrane surface is apparently not mediated by the concave interface displayed by a dimer of FBAR domains but rather by its lateral surface. Intermolecular interactions between neighboring dimers (tip-to-tip and lateral) are formed upon oligomerization. This generates a continuous helical coat, in which individual FBAR domains are rearranged such that their concave faces contact the membrane surface (Fig. 3). Most of the identified FBAR proteins, including Tocas/FBP17/CIP4, and syndapin contain SH3 domains that interact with dynamin and synaptojanin, and in select cases also with N-WASP. FBAR proteins may thus tighten the link between dynamin function at the vesicle neck and re-arrangements of the actin cytoskeleton (Roux et al. 2006) as further discussed below.

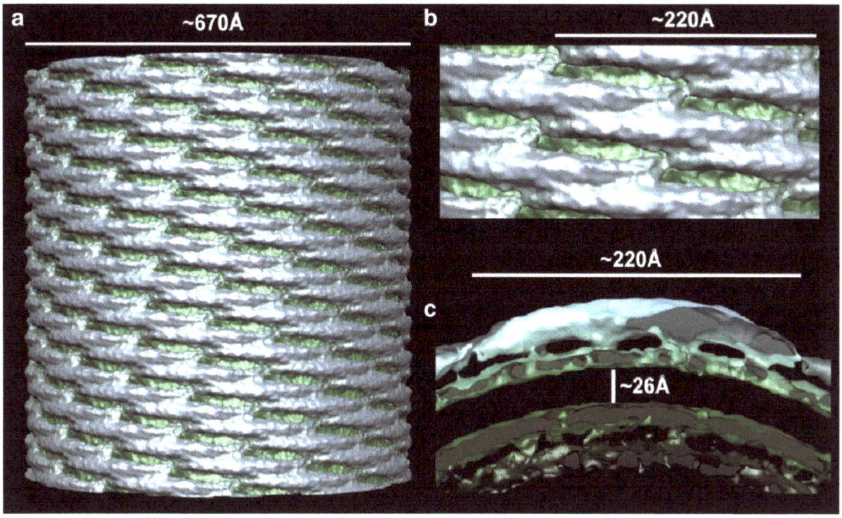

Fig. 3 a–c Some F-BAR proteins can induce membrane curvature: single particle helical reconstruction of a CIP4 F-BAR domain-induced membrane tubule. **a** Surface of a >67-nm diameter membrane tubule at >17-Å resolution. The protein coat is colored *blue-gray* and the underlying membrane is *green*. **b** Zoom in on the lattice seen orthogonal to the cylindrical axis, highlighting the tip-to-tip interactions and the broad contacts between laterally adjacent dimers. **c** Cross-sectional slab through one dimer parallel with the plane of the tip-to-tip interaction. There are four clearly resolved points of membrane binding. The hydrophobic core of the phospholipid bilayer is >26 Å thick and the headgroup regions are >12 Å thick. (Reprinted from Frost et al. 2008, with permission from Elsevier)

Assisting Curvature Induction: Local Changes in Lipid Composition and Organization

So far we have only considered proteins driving membrane deformation at the plasma membrane. However, a major class of membrane constituents are lipids that indeed play fundamental roles in the regulation of endocytic processes. One class of phospholipids that have drawn special attention are PIs, differentially phosphorylated, short-lived derivatives of phosphatidylinositol. Based on their distinctive subcellular distribution and binding characteristics they have been suggested to constitute spatial landmarks which recruit proteins harboring appropriate PI-binding domains including many membrane-deforming proteins (Behnia and Munro 2005). PIs are highly interconvertible by appropriate PI kinases and phosphatases, a property that renders them especially suited for regulating intracellular membrane dynamics. Their turnover can trigger changes in binding characteristics of individual membranes and constitutes a means to rapidly shift membrane identity. It is generally assumed that PIs are not equally distributed within the cytoplasmic leaflet of organellar membranes but synthesized de novo at discrete sites. Thus, whereas the total concentrations of individual PIs in a living cell are considered to be relatively low, local concentrations might rise to levels high enough to recruit corresponding cohorts of PI-binding and membrane-bending proteins (Fig. 2a, b). How PI metabolism and the generation of PI-hotspots are regulated is still largely unknown (Krauss and Haucke 2007).

The prevalent PI species within the plasma membrane is PI(4,5)P_2. Its synthesis is mediated predominantly by phosphatidylinositol 4-phosphate 5-kinases (type I PIPKs). PI(4,5)P_2 is required for both clathrin-dependent and -independent pathways of endocytosis. PI(4,5)P_2 has also been implicated in regulating actin dynamics (DiPaolo and de Camilli 2006) at the cell surface. Therefore, its synthesis has been linked to both clathrin-dependent and -independent endocytic pathways. This is best illustrated by experiments in which a 5-phosphatase has been targeted to the plasmalemma either constitutively or following acute rapamycin-mediated membrane recruitment. In both cases selective loss of PI(4,5)P_2 has been shown to completely inhibit internalization of the transferrin and epidermal growth factor receptors (Krauss et al. 2003; Varnai et al. 2006) due to dissociation of clathrin and AP-2 from the plasma membrane (Zoncu et al. 2007).

How is the synthesis of individual subpools of PI(4,5)P_2 regulated? Several studies indicate that activated small GTPases are capable of stimulating type I PIPKs. Among these are Rho, Rac, and Cdc42 that regulate PI(4,5)P_2-mediated rearrangements of the actin cytoskeleton (Santarius et al. 2006). Similarly, the small GTPase ARF6 has been implicated in PI(4,5)P_2-dependent trafficking processes at the plasma membrane (Donaldson 2005). ARF6-GTP directly associates with type I PIPKs (Honda et al. 1999), and via this interaction can enhance recruitment of clathrin/AP-2 coats to synaptic vesicle membranes (Krauss et al. 2003). In many cases PIPK I activation is mediated by cooperative mechanisms: Arf1 and Arf6 also stimulate phospholipase D (PLD). The resulting product, phosphatidic acid, significantly enhances PIPK I-mediated PI(4,5)P_2 synthesis (Powner and Wakelam 2002). Conversely, PI(4,5)P_2 greatly stimulates PLD activity. In concert these mechanisms create a positive feed-forward loop resulting in high local concentrations of phosphatidic acid and PI(4,5)P_2 at sites of Arf activation. Arf activity in turn is under tight regulatory control by guanine nucleotide exchange factors and GTPase-activating proteins, some of which have been functionally linked with actin remodeling and distinct endocytic pathways (Casanova 2007; Gillingham and Munro 2007). Interestingly, some of these GTPase-activating proteins and guanine nucleotide exchange factors contain PI-binding domains that guide them to membranes, i.e., upon growth factor signaling, thereby creating an additional layer of spatiotemporal control.

A pool of PI(4,5)P_2 specifically dedicated to CME appears to be regulated by AP-2. AP-2 associates with type I PIPKs, in particular its γ-isoform, via up to three distinct binding sites (Bairstow et al. 2006; Krauss et al. 2006; Nakano-Kobayashi et al. 2007). Interestingly, kinase activity is stimulated upon concomitant binding of AP-2μ to YxxF-based endocytic motifs. PI(4,5)P_2 in turn is required for the association of AP-2 with endocytic motifs, resulting in a coincidence detection mechanism whereby coated pit formation is directed to sites of high local concentrations of PI(4,5)P_2 and cargo membrane proteins (Honing et al. 2005).

Given the metabolic interplay between formation of PI(4,5)P_2 and phosphatidic acid one might hypothesize that PI binding protein-mediated membrane deformation is coupled to local changes in the distribution of membrane lipids between the two leaflets of the bilayer. Lysophosphatidic acid and phosphatidic acid are interconverted by lysophosphatidic acid acyl transferase and phospholipase A_2. Due to their distinct geometries lysophosphatidic acid and phosphatidic acid promote opposite curvatures (Kooijman et al. 2005). Thus, localized activation of either enzyme might support or inhibit the formation of membrane invaginations during endocytosis. Furthermore, lipids display headgroups of distinct sizes and will thus occupy different membrane areas in opposing leaflets. Hence, asymmetric lipid distribution likely affects membrane curvature. It is worth noting that flippases which promote lipid transfer from one leaflet of the membrane to the other regulate CME of transferrin in mammalian cells (Farge et al. 1999) and internalization of both receptors and fluid phase markers in yeast (Pomorski et al. 2003).

Recent data indicate that lateral sequestration of DRM-associated lipids is required for uptake of Shiga toxin (STx) via induction of tubular membrane invaginations (Romer et al. 2007). The B-subunit of the bacterial STx enters cells in a clathrin-independent manner involving its association with globotriaosyl ceramide (Gb3) at the cell surface. STxB induces lipid reorganization that favors negative membrane curvature resulting in the formation of narrow tubular invaginations (Fig. 2d). Interestingly, STxB-induced tubulation was enhanced upon energy depletion and inhibition of dynamin or actin function, suggesting that dynamin plays a key role in scission, but not in tubule formation. Even more striking was the finding that the B-subunit of STx can induce similar invaginations in model membranes of appropriate composition, e.g., in the absence of other cellular factors. Given that the protein forms homopentamers and that each subunit can bind up to three Gb3 molecules, binding of the toxin to its receptor might induce the formation of Gb3-STxB-clusters that have the propensity to induce negatively curved membrane microdomains. How exactly microdomain expansion and tubule elongation are linked remains an open question.

A Matter of Stretching and Squeezing: Mechanical Forces Sculpt the Membrane

Endocytic events are intimately linked with remodeling of the actin cytoskeleton. Actin is frequently observed at sites of high membrane curvature, and has therefore been suggested to function during macropinocytosis, phagocytosis, and CME (Fig. 1). Its polymerization could generate forces that support the fission step (Yarar et al. 2005). In line with this possibility, dual-color experiments applying TIRF microscopy have revealed that during CME, clathrin/AP-2 are recruited to sites of cargo entry prior to dynamin. At late stages actin regulators including cortactin, N-WASP, Arp2/3, and actin itself appear at endocytic sites. The formation of F-actin might push the vesicle away from the plasma membrane and create further tension to the vesicle neck until fission is mediated by dynamin through a conformational change induced by GTP hydrolysis.

Motor proteins associating with the local actin network and the forming vesicle or tubule might aid constriction of the neck and final abscission (Soldati and Schliwa 2006). Dynamin and synaptojanin can interact with the plus-end-directed motor myosin IE through their PRDs (Krendel et al. 2007), whereas the adaptor protein Dab2 together with PI(4,5)P_2 can recruit the minus-end-directed motor myosin VI to clathrin-coated pits (Dance et al. 2004; Spudich et al. 2007). Opposite motor activities attached to the vesicle and the stalk might then aid abscission and/or constriction of the neck (Soldati and Schliwa 2006).

Actin regulatory proteins initiate actin polymerization at sites enriched in PI(4,5)P_2 (Logan and Mandato 2006). Since many clathrin-dependent and -independent endocytic processes are associated with the generation of local PI(4,5)P_2-pools, this might represent a common mechanism to support expansion of curved microdomains. The force-generating ability of actin is well illustrated by actin-comet-like vesicles that are propelled through the cytoplasm upon overexpression of type I PIPKs, of constitutively active Arf6, or of dynamin 2 in fibroblasts (Orth et al. 2002; Schafer et al. 2000). In line with a general role for actin in different modes of endocytosis, SNX9 has recently been shown to localize to clathrin-coated pits and to actin-rich structures implicated in fluid-phase uptake and internalization of GPI-anchored proteins (Yarar et al. 2005). Depletion of SNX9 significantly impaired fluid-phase uptake, but had no effect on a compensatory fluid-phase internalization pathway initiated by overexpression of an inactive dynamin mutant. This indicates that SNX9 function is restricted to select endocytic pathways. In vitro SNX9 has been shown to enhance actin polymerization stimulated by Arp2/3 and N-WASP. This effect was even more pronounced in the presence of $PI(4,5)P_2$-enriched liposomes. Thus, SNX9 might fulfill similar functions during clathrin-dependent and -independent endocytosis. These include sensing of $PI(4,5)P_2$-enriched membrane surfaces and coordination of curvature acquisition with rearrangements of the actin cytoskeleton.

Outlook

Whereas our knowledge about curvature-generating mechanisms during CME and phagocytosis is quite extensive, mechanisms that drive invagination of the plasma membrane during clathrin-independent endocytosis are still poorly understood. A variety of entry pathways have been described to date, but there might be even more complexity in different cellular systems than previously anticipated. Pathways used by a particular cargo may differ depending on cell type, the cellular conditions, or the subcellular compartment to which they are targeted.

In some cases binding of extracellular components or particles to the cell surface must be transmitted across the membrane to trigger appropriate intracellular responses and favor membrane bending. How this is achieved when no transmembrane receptors are involved is still unclear. Recent evidence on STx entry into cells (Romer et al. 2007) indicates that clustering of lipid molecules induced by toxin binding apparently provides sufficient energy for initial membrane bending. Similar mechanisms might also apply for other clathrin-independent pathways, at least where cargoes associate preferentially with distinct lipid species. Pre-curved membranes might become targets for curvature-sensing molecules such as BAR domain-containing proteins which then provide a link to actin regulatory proteins or dynamin. Further mechanisms that sense differential lipid packing most likely contribute to this process. A motif termed ALPS (Table 1) that may fulfill the role of a lipid-packing sensor by inserting into bent membranes has been identified in ArfGAP1, a protein involved in the formation of carrier vesicles from Golgi membranes (Drin et al. 2007). The ALPS motif comprises an amphipathic α-helix which serves to couple ArfGAP1 activity to curved membrane buds stabilized by COPI coat proteins (Bigay et al. 2003). It is tempting to speculate that factors with similar properties might function at the plasma membrane as well, and could thereby control the activities of small GTPases involved in distinct endocytic events.

Acknowledgements: The authors acknowledge support from the Deutsche Forschungsgemeinschaft (DFG) (HA2686/3-1/FOR806; HA2686/4-1/SFB740).

References

Antonny B, Beraud-Dufour S, Chardin P, Chabre M (1997) N-terminal hydrophobic residues of the G-protein ADP-ribosylation factor-1 insert into membrane phospholipids upon GDP to GTP exchange. Biochemistry 36:4675–4684

Bairstow SF, Ling K, Su X, Firestone AJ, Carbonara C, Anderson RA (2006) Type Igamma661 phosphatidylinositol phosphate kinase directly interacts with AP2 and regulates endocytosis. J Biol Chem 281:20632–20642

Beck R, Sun Z, Adolf F, Rutz C, Bassler J, Wild K, Sinning I, Hurt E, Brügger B, Béthune J, Wieland F (2008) Membrane curvature induced by Arf1-GTP is essential for vesicle formation. Proc Natl Acad Sci U S A 105:11731–11736

Behnia R, Munro S (2005) Organelle identity and the signposts for membrane traffic. Nature 438:597–604

Bigay J, Gounon P, Robineau S, Antonny B (2003) Lipid packing sensed by ArfGAP1 couples COPI coat disassembly to membrane bilayer curvature. Nature 426:563–566

Bonazzi M, Cossart P (2006) Bacterial entry into cells: a role for the endocytic machinery. FEBS Lett 580:2962–2967

Bonifacino JS, Traub LM (2003) Signals for sorting of transmembrane proteins to endosomes and lysosomes. Annu Rev Biochem 72:395–447

Casanova JE (2007) Regulation of Arf activation: the Sec7 family of guanine nucleotide exchange factors. Traffic 8:1476–1485

Chimini G, Chavrier P (2000) Function of Rho family proteins in actin dynamics during phagocytosis and engulfment. Nat Cell Biol 2:E191–196

Cohen AW, Hnasko R, Schubert W, Lisanti MP (2004) Role of caveolae and caveolins in health and disease. Physiol Rev 84:1341–1379

Damm EM, Pelkmans L, Kartenbeck J, Mezzacasa A, Kurzchalia T, Helenius A (2005) Clathrin- and caveolin-1-independent endocytosis: entry of simian virus 40 into cells devoid of caveolae. J Cell Biol 168:477–488

Dance AL, Miller M, Seragaki S, Aryal P, White B, Aschenbrenner L, Hasson T (2004) Regulation of myosin-VI targeting to endocytic compartments. Traffic 5:798–813

Delaney KA, Murph MM, Brown LM, Radhakrishna H (2002) Transfer of M2 muscarinic acetylcholine receptors to clathrin-derived early endosomes following clathrin-independent endocytosis. J Biol Chem 277:33439–33446

Di Paolo G, De Camilli P (2006) Phosphoinositides in cell regulation and membrane dynamics. Nature 443:651–657

Donaldson JG (2003) Multiple roles for Arf6: sorting, structuring, and signaling at the plasma membrane. J Biol Chem 278:41573–41576

Donaldson JG (2005) Arfs, phosphoinositides and membrane traffic. Biochem Soc Trans 33:1276–1278

Drab M, Verkade P, Elger M et al (2001) Loss of caveolae, vascular dysfunction, and pulmonary defects in caveolin-1 gene-disrupted mice. Science 293:2449–2452

Drin G, Casella JF, Gautier R, Boehmer T, Schwartz TU, Antonny B (2007) A general amphipathic alpha-helical motif for sensing membrane curvature. Nat Struct Mol Biol 14:138–146

Edeling MA, Smith C, Owen D (2006) Life of a clathrin coat: insights from clathrin and AP structures. Nat Rev Mol Cell Biol 7:32–44

Edidin M (2001) Shrinking patches and slippery rafts: scales of domains in the plasma membrane. Trends Cell Biol 11:492–496

Farge E, Ojcius DM, Subtil A, Dautry-Varsat A (1999) Enhancement of endocytosis due to aminophospholipid transport across the plasma membrane of living cells. Am J Physiol 276:C725–733

Ford MG, Mills IG, Peter BJ, Vallis Y, Praefcke GJ, Evans PR, McMahon HT (2002) Curvature of clathrin-coated pits driven by epsin. Nature 419:361–366

Fotin A, Cheng Y, Grigorieff N, Walz T, Harrison SC, Kirchhausen T (2004a) Structure of an auxilin-bound clathrin coat and its implications for the mechanism of uncoating. Nature 432:649–653

Fotin A, Cheng Y, Sliz P, Grigorieff N, Harrison SC, Kirchhausen T, Walz T (2004b) Molecular model for a complete clathrin lattice from electron cryomicroscopy. Nature 432:573–579

Frick M, Bright NA, Riento K, Bray A, Merrified C, Nichols BJ (2007) Coassembly of flotillins induces formation of membrane microdomains, membrane curvature, and vesicle budding. Curr Biol 17:1151–1156

Frost A, Perera R, Roux A, Spasov K, Destaing O, Egelman EH, De Camilli P, Unger VM (2008) Structural basis of membrane invagination by F-BAR domains. Cell 132:807–817
Gad H, Ringstad N, Löw P et al (2000) Fission and uncoating of synaptic clathrin-coated vesicles are perturbed by disruption of interactions with the SH3 domain of endophilin. Neuron 27:301–312
Gallop JL, Jao CC, Kent HM, Butler PJ, Evans PR, Langen R, McMahon HT (2006) Mechanisms of endophilin N-BAR domain-mediated membrane curvature. EMBO J 25:2898–2910
Gillingham AK, Munro S (2007) The small G proteins of the Arf family and their regulators. Annu Rev Cell Dev Biol 23:579–611
Glebov OO, Bright NA, Nichols BJ (2006) Flotillin-1 defines a clathrin-independent endocytic pathway in mammalian cells. Nat Cell Biol 8:46–54
Godi A, Di Campli A, Konstantakopoulos A et al (2004) FAPPs control Golgi-to-cell-surface membrane traffic by binding to ARF and PtdIns(4)P. Nat Cell Biol 6:393–404
Gold ES, Underhill DM, Morrissette NS, Guo J, McNiven MA, Aderem A (1999) Dynamin 2 is required for phagocytosis in macrophages. J Exp Med 190:1849–1856
Grabs D, Slepnev VI, Songyang Z, David C, Lynch M, Cantley LC, De Camilli P (1997) The SH3 domain of amphiphysin binds the proline-rich domain of dynamin at a single site that defines a new SH3 binding consensus sequence. J Biol Chem 272:13419–13425
Grass I, Thiel S, Honing S, Haucke V (2004) Recognition of a basic AP-2 binding motif within the C2B domain of synaptotagmin is dependent on multimerization. J Biol Chem 279:54872–54880
Habermann B (2004) The BAR-domain family of proteins: a case of bending and binding? EMBO Rep 5:250–255
Harder T (2004) Lipid raft domains and protein networks in T-cell receptor signal transduction. Curr Opin Immunol 16:353–359
Haucke V (2005) Phosphoinositide regulation of clathrin-mediated endocytosis. Biochem Soc Trans 33:1285–1289
Heuser J (1980) Three-dimensional visualization of coated vesicle formation in fibroblasts. J Cell Biol 84:560–583
de Heuvel E, Bell AW, Ramjaun AR, Wong K, Sossin WS, McPherson PS (1997) Identification of the major synaptojanin-binding proteins in brain. J Biol Chem 272:8710–8716
Hicke L, Dunn R (2003) Regulation of membrane protein transport by ubiquitin and ubiquitin-binding proteins. Annu Rev Cell Dev Biol 19:141–172
Hinrichsen L, Meyerholz A, Groos S, Ungewickell EJ (2006) Bending a membrane: how clathrin affects budding. Proc Natl Acad Sci USA 103:8715–8720
Hinshaw JE (2000) Dynamin and its role in membrane fission. Annu Rev Cell Dev Biol 16:483–519
Honda A, Nogami M, Yokozeki T et al (1999) Phosphatidylinositol 4-phosphate 5-kinase alpha is a downstream effector of the small G protein ARF6 in membrane ruffle formation. Cell 99:521–532
Honing S, Ricotta D, Krauss M et al (2005) Phosphatidylinositol-(4,5)-bisphosphate regulates sorting signal recognition by the clathrin-associated adaptor complex AP2. Mol Cell 18:519–531
Huynh KK, Kay JG, Stow JL, Grinstein S (2007) Fusion, fission, and secretion during phagocytosis. Physiology (Bethesda) 22:366–372
Itoh T, De Camilli P (2006) BAR, F-BAR (EFC) and ENTH/ANTH domains in the regulation of membrane-cytosol interfaces and membrane curvature. Biochim Biophys Acta 1761:897–912
Jung N,Wienisch M, Gu M et al (2007) Molecular basis of synaptic vesicle cargo recognition by the endocytic sorting adaptor stonin 2. J Cell Biol 179:1497–1510
Keyel PA, Mishra SK, Roth, R, Heuser JE, Watkins SC, Traub LM (2006) A single common portal for clathrin-mediated endocytosis of distinct cargo governed by cargo-selective adaptors. Mol Biol Cell 17:4300–4317
Kirkham M, Fujita A, Chadda R et al (2005) Ultrastructural identification of uncoated caveolin-independent early endocytic vehicles. J Cell Biol 168:465–476
Kooijman EE, Chupin V, Fuller NL, Kozlov MM, de Kruijff B, Burger KN, Rand PR (2005) Spontaneous curvature of phosphatidic acid and lysophosphatidic acid. Biochemistry 44:2097–2102

Krendel M, Osterweil EK, Mooseker MS (2007) Myosin 1E interacts with synaptojanin-1 and dynamin and is involved in endocytosis. FEBS Lett 581:644–650

Krauss M, Haucke V (2007) Phosphoinositide-metabolizing enzymes at the interface between membrane traffic and cell signalling. EMBO Rep 8:241–246

Krauss M, Kinuta M, Wenk MR, De Camilli P, Takei K, Haucke V (2003) ARF6 stimulates clathrin/AP-2 recruitment to synaptic membranes by activating phosphatidylinositol phosphate kinase type Igamma. J Cell Biol 162:113–124

Krauss M, Kukhtina V, Pechstein A, Haucke V (2006) Stimulation of phosphatidylinositol kinase type I-mediated phosphatidylinositol (4,5)-bisphosphate synthesis by AP-2mu-cargo complexes. Proc Natl Acad Sci USA 103:11934–11939

Krauss M, Jia JY, Roux A, Beck R, Wieland FT, De Camili P, Haucke V (2008) Arf1-GTP-induced tubule formation suggests a function of Arf family proteins in curvature acquisition at sites of vesicle budding. J Biol Chem 283:27717–27723

Kumari S, Mayor S (2008) ARF1 is directly involved in dynamin-independent endocytosis. Nat Cell Biol 10:30–41

Lakadamyali M, Rust MJ, Zhuang X (2006) Ligands for clathrin-mediated endocytosis are differentially sorted into distinct populations of early endosomes. Cell 124:997–1009

Lamaze C, Dujeancourt A, Baba T, Lo CG, Benmerah A, Dautry-Varsat A (2001) Interleukin 2 receptors and detergent-resistant membrane domains define a clathrin-independent endocytic pathway. Mol Cell 7:661–671

Le PU, Guay G, Altschuler Y, Nabi IR (2002) Caveolin-1 is a negative regulator of caveolae-mediated endocytosis to the endoplasmic reticulum. J Biol Chem 277:3371–3379

Lecuit M, Hurme R, Pizarro-Cerda J, Ohayon H, Geiger B, Cossart P (2000) A role for alpha-and beta-catenins in bacterial uptake. Proc Natl Acad Sci USA 97, 10008–10013

Lencer WI, Hirst TR, Holmes RK (1999) Membrane traffic and the cellular uptake of cholera toxin. Biochim Biophys Acta 1450:177–190

Lundmark R, Carlsson SR (2002) The beta-appendages of the four adaptor-protein (AP) complexes: structure and binding properties, and identification of sorting nexin 9 as an accessory protein to AP-2. Biochem J 362:597–607

Lundmark R, Carlsson SR (2003) Sorting nexin 9 participates in clathrin-mediated endocytosis through interactions with the core components. J Biol Chem 278:46772–46781

Lundmark R, Carlsson SR (2004) Regulated membrane recruitment of dynamin-2 mediated by sorting nexin 9. J Biol Chem 279:42694–42702

Masuda M, Takeda S, Sone M, Ohki T, Mori H, Kamioka Y, Mochizuki N (2006) Endophilin BAR domain drives membrane curvature by two newly identified structure-based mechanisms. EMBO J 25:2889–2897

May RC, Machesky LM (2001) Phagocytosis and the actin cytoskeleton. J Cell Sci 114:1061–1077

Mayor S, Pagano RE (2007) Pathways of clathrin-independent endocytosis. Nat Rev Mol Cell Biol 8:603–612

McMahon HT, Gallop JL (2005) Membrane curvature and mechanisms of dynamic cell membrane remodelling. Nature 438:590–596

Monier S, Parton RG, Vogel F, Behlke J, Henske A, Kurzchalia TV (1995) VIP21-caveolin, a membrane protein constituent of the caveolar coat, oligomerizes in vivo and in vitro. Mol Biol Cell 6:911–927

Musacchio A, Smith CJ, Roseman AM, Harrison SC, Kirchhausen T, Pearse BM (1999) Functional organization of clathrin in coats: combining electron cryomicroscopy and X-ray crystallography. Mol Cell 3:761–770

Myers KR, Casanova JE (2008) Regulation of actin cytoskeleton dynamics by Arf-family GTPases. Trends Cell Biol 18:184–192

Nakano-Kobayashi A, Yamazaki M, Unoki T et al (2007) Role of activation of PIP5Kgamma661 by AP-2 complex in synaptic vesicle endocytosis. EMBO J 26:1105–1116

Naslavsky N, Weigert R, Donaldson JG (2004) Characterization of a nonclathrin endocytic pathway: membrane cargo and lipid requirements. Mol Biol Cell 15:3542–3552

Niedergang F, Chavrier P (2004) Signaling and membrane dynamics during phagocytosis: many roads lead to the phagos(R)ome. Curr Opin Cell Biol 16:422–428
Orth JD, McNiven MA (2003) Dynamin at the actin-membrane interface. Curr Opin Cell Biol 15:31–39
Orth JD, Krueger EW, Cao H, McNiven MA (2002) The large GTPase dynamin regulates actin comet formation and movement in living cells. Proc Natl Acad Sci USA 99:167–172
Palade GE, Bruns RR (1968) Structural modulations of plasmalemmal vesicles. J Cell Biol 37:633–649
Pasqualato S, Menetrey J, Franco M, Cherfils J (2001) The structural GDP/GTP cycle of human Arf6. EMBO Rep 2:234–238
Pelkmans L, Helenius A (2002) Endocytosis via caveolae. Traffic 3:311–320
Pelkmans L, Zerial M (2005) Kinase-regulated quantal assemblies and kiss-and-run recycling of caveolae. Nature 436:128–133
Pelkmans L, Kartenbeck J, Helenius A (2001) Caveolar endocytosis of simian virus 40 reveals a new two-step vesicular-transport pathway to the ER. Nat Cell Biol 3:473–483
Peter BJ, Kent HM, Mills IG, Vallis Y, Butler PJ, Evans PR, McMahon HT (2004) BAR domains as sensors of membrane curvature: the amphiphysin BAR structure. Science 303:495–499
Pomorski T, Lombardi R, Riezman H, Devaux PF, van Meer G, Holthuis JC (2003) Drs2p-related P-type ATPases Dnf1p and Dnf2p are required for phospholipid translocation across the yeast plasma membrane and serve a role in endocytosis. Mol Biol Cell 14:1240–1254
Powner DJ, Wakelam MJ (2002) The regulation of phospholipase D by inositol phospholipids and small GTPases. FEBS Lett 531:62–64
Presley JF, Ward TH, Pfeifer AC, Siggia ED, Phair RD, Lippincott-Schwartz J (2002) Dissection of COPI and Arf1 dynamics in vivo and role in Golgi membrane transport. Nature 417:187–193
Puthenveedu MA, von Zastrow M (2006) Cargo regulates clathrin-coated pit dynamics. Cell 127:113–124
Pylypenko O, Lundmark R, Rasmuson E, Carlsson SR, Rak A (2007) The PX-BAR membrane-remodeling unit of sorting nexin 9. EMBO J 26:4788–4800
Radhakrishna H, Klausner RD, Donaldson JG (1996) Aluminum fluoride stimulates surface protrusions in cells overexpressing the ARF6 GTPase. J Cell Biol 134:935–947
Razani B, Woodman SE, Lisanti MP (2002) Caveolae: from cell biology to animal physiology. Pharmacol Rev 54:431–467
Ringstad N, Gad H, Low P, Di Paolo G, Brodin L, Shupliakov O, De Camilli P (1999) Endophilin/SH3p4 is required for the transition from early to late stages in clathrin-mediated synaptic vesicle endocytosis. Neuron 24:143–154
Romer W, Berland L, Chambon V et al (2007) Shiga toxin induces tubular membrane invaginations for its uptake into cells. Nature 450:670–675
Rothberg KG, Heuser JE, Donzell WC, Ying YS, Glenney JR, Anderson RG (1992) Caveolin, a protein component of caveolae membrane coats. Cell 68:673–682
Roux A, Uyhazi K, Frost A, De Camilli P (2006) GTP-dependent twisting of dynamin implicates constriction and tension in membrane fission. Nature 441:528–531
Sabharanjak S, Sharma P, Parton RG, Mayor S (2002) GPI-anchored proteins are delivered to recycling endosomes via a distinct cdc42-regulated, clathrin-independent pinocytic pathway. Dev Cell 2:411–423
Santarius M, Lee CH, Anderson RA (2006) Supervised membrane swimming: small G-protein lifeguards regulate PIPK signalling and monitor intracellular PtdIns(4,5)P2 pools. Biochem J 398:1–13
Schafer DA, D'Souza-Schorey C, Cooper JA (2000) Actin assembly at membranes controlled by ARF6. Traffic 1:892–903
Schuske KR, Richmond JE, Matthies DS, Davis WS, Runz S, Rube DA, van der Bliek AM, Jorgensen EM (2003) Endophilin is required for synaptic vesicle endocytosis by localizing synaptojanin. Neuron 40:749–762
Shimada A, Niwa H, Tsujita K et al (2007) Curved EFC/F-BAR-domain dimers are joined end to end into a filament for membrane invagination in endocytosis. Cell 129:761–772

Shin N, Ahn N, Chang-Ileto B et al (2008) SNX9 regulates tubular invagination of the plasma membrane through interaction with actin cytoskeleton and dynamin 2. J Cell Sci 121:1252–1263
Simons K, Ikonen E (1997) Functional rafts in cell membranes. Nature 387:569–572
Slepnev VI, Ochoa GC, Butler MH, De Camilli P (2000) Tandem arrangement of the clathrin and AP-2 binding domains in amphiphysin 1 and disruption of clathrin coat function by amphiphysin fragments comprising these sites. J Biol Chem 275:17583–17589
Soldati T, Schliwa M (2006) Powering membrane traffic in endocytosis and recycling. Nat Rev Mol Cell Biol 7:897–908
Spudich G, Chibalina MV, Au JS, Arden SD, Buss F, Kendrick-Jones J (2007) Myosin VI targeting to clathrin-coated structures and dimerization is mediated by binding to Disabled-2 and PtdIns(4,5)P2. Nat Cell Biol 9:176–183
Takei K, Slepnev VI, Haucke V, De Camilli P (1999) Functional partnership between amphiphysin and dynamin in clathrin-mediated endocytosis. Nat Cell Biol 1:33–39
Thomsen P, Roepstorff K, Stahlhut M, van Deurs B (2002) Caveolae are highly immobile plasma membrane microdomains, which are not involved in constitutive endocytic trafficking. Mol Biol Cell 13:238–250
Varnai P, Thyagarajan B, Rohacs T, Balla T (2006) Rapidly inducible changes in phosphatidylinositol 4,5-bisphosphate levels influence multiple regulatory functions of the lipid in intact living cells. J Cell Biol 175:377–382
Veiga E, Cossart P (2005) Listeria hijacks the clathrin-dependent endocytic machinery to invade mammalian cells. Nat Cell Biol 7:894–900
Verstreken P, Koh TW, Schulze KL et al (2003) Synaptojanin is recruited by endophilin to promote synaptic vesicle uncoating. Neuron 40:733–748
Vicente-Manzanares M, Sanchez-Madrid F (2004) Role of the cytoskeleton during leukocyte responses. Nat Rev Immunol 4:110–122
Walseng E, Bakke O, Roche PA (2008) Major histocompatibility complex class II-peptide complexes internalize using a clathrin- and dynamin-independent endocytosis pathway. J Biol Chem 283:14717–14727
Wolfe BL, Trejo J (2007) Clathrin-dependent mechanisms of G protein-coupled receptor endocytosis. Traffic 8:462–470
Yarar D, Waterman-Storer CM, Schmid SL (2005) A dynamic actin cytoskeleton functions at multiple stages of clathrin-mediated endocytosis. Mol Biol Cell 16:964–975
Zoncu R, Perera RM, Sebastian R et al (2007) Loss of endocytic clathrin-coated pits upon acute depletion of phosphatidylinositol 4,5-bisphosphate. Proc Natl Acad Sci USA 104:3793–3798

The Long Journey: Actin on the Road to Pro- and Eukaryotic Cells

Brigitte M. Jockusch and Peter L. Graumann

Abstract Actin-like proteins comprise a large group of polymorphic proteins that readily form filaments engaged in cytoskeletal functions. Various members have been identified in prokaryotic and eukaryotic cells, e.g. MreB, ParM and Ta0583, and actin and the actin-related proteins, ARPs, respectively. Therefore, it is assumed that an ancestor of actin/MreB/ParM already existed in the last common progenitor of all cells. In eubacteria and archaea, actin-like proteins are either membrane-associated or freely soluble, and their activities are related to motility, cell shape maintenance, subcellular organization and cell cycle progression. In eukaryotes, all these functions are executed by actin in various isoforms. Additional functions have been described for actin and ARPs in the nucleus of the eukaryotic cell, and some of those were also discovered in prokaryotes. In the current essay, we compare structures and selected functions of prokaryotic and eukaryotic actins and discuss various aspects on how actins may have found their way into bacteria, into the eukaryotic cytoplasm and into the nuclear compartment.

1 Actin Homologues Are Present in Pro- and Eukaryotes

Actin has been discovered more than sixty years ago in muscle extracts, when biochemical methods were already sophisticated and highly reliable (reviewed in (Schleicher and Jockusch 2008)). Since then, it has become one of the most intensively studied proteins. Being very abundant in most eukaryotic cells and based on

---

Dedicated to Jürgen Wehland, who shared our passion for the actin cytoskeleton

B.M. Jockusch (✉)
Cell Biology, Zoological Institute, Technical University of Braunschweig, 38092 Braunschweig, Germany
e-mail: bmj@tu-bs.de

Rev Physiol Biochem Pharmacol, DOI: 10.1007/112_2011_1,

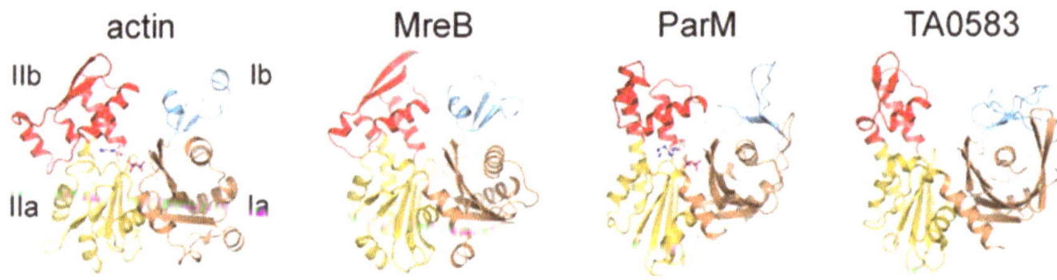

Fig. 1 Structural homologies of actin-like proteins in pro-and eukaryotic organisms. The overall arrangement of the sequences in two large domains (I and II), comprising two subdomains each (Ia and Ib and IIa and IIb), is seen in the atomic model of muscle actin (Kabsch et al. 1990). The nucleotide is bound in the deep cleft between I and II. The structures of the bacterial proteins ParM (*E. coli*), MreB (*T. maritima*) and the archaean protein Ta0583 (from *Thermoplasma acidophilum*) show a striking similarity to the mammalian actin, as demonstrated by the colouring of all four subdomains ((Roeben et al. 2006), courtesy of Andreas Bracher, Max-Planck-Institute for Biochemistry, Munich, Germany)

cell- and tissue-specific expression of several isoforms., eukaryotic actin is a key player in all processes involving motility.

The identification of the prokaryotic proteins MreB and ParM as homologues of actin had to await the area of molecular genetics, combined with refined light and electron microscopy (reviewed in (Graumann 2007)). Cells lacking MreB or expressing mutants thereof were found to be round (coccoid) instead of rod shaped, but the relation to actin was first hinted at in a landmark paper in 1992, when the “actin superfamily” was coined according to conserved sequences encompassing the ATP binding pocket, and contained actin as well as sugar kinases, HSP70 chaperones, bacterial MreB and StbA (now called ParM). In their publication (Bork et al. 1992), Bork, Sander and Valencia noted MreB, StbA and actin as most closely related, which was completely supported by the studies to come. A comparison of the structural details of the prokaryotic proteins MreB and ParM (Van den Ent et al. 2001) with that of muscle actin (Kabsch et al. 1990) demonstrates the striking similarity of their morphology. All molecules are organised in two large domains (Fig. 1, indicated in the actin molecule as I and II), each comprising two subdomains (Fig. 1, Ia, Ib and IIa, IIb), with the same secondary folds, and harbouring a nucleotide-binding pocket in the cleft between I and II. Furthermore, the analysis of the formation and subunit position of actin, ParM and MreB in polar filaments (Van den Ent et al. 2002) left no doubt that the proteins evolved from an ancestor that was able to polymerise in the presence of ATP (or GTP) and magnesium. MreB forms bundles of filaments that run underneath the membrane in a helical path in many bacteria, and it is essential for the maintenance of proper cell morphology (Figge et al. 2004; Jones et al. 2001; Kruse et al. 2003). In many bacteria, MreB filaments are highly dynamic and remodel within few minutes (Carballido-Lopez and Errington 2003; Defeu Soufou and Graumann 2004; Kim et al. 2006), which is important for their function. Electron microscopy has demonstrated that protofilaments formed by MreB are similar to the long-pitch helical strand of the double stranded F-actin, displaying contacts at more or less the same sites, with subunit spacings of 51 or 55, respectively (Van den Ent et al. 2001). While double stranded protofilaments of MreB are straight, ParM forms double stranded filaments quite similar to muscle actin. Notably, different states of handedness for ParM protofilaments have been identified by crystallography, indicating that this protein can undergo substantial domain–domain motions (Orlowa et al. 2007; Popp et al. 2008; Van den Ent et al. 2002), quite comparable to those observed recently for muscle actin filaments (Oda et al. 2009). For MreB, it has been shown that this protein can also form multilayered sheets consisting of interwoven filaments (Popp et al. 2010). Thus, at least some of the prokaryotic actins are polymorphic proteins, capable to form not only filaments, but a variety of supramolecular structures. The structural plasticity of eukaryotic actins in the context of different tasks and their location in different cellular compartments has been discussed in (Jockusch et al. 2006). Supramolecular assemblies of these proteins can be visualized in every modern pro- or eukaryotic cell (see Fig. 2).

Over the past decade, the catalogue of actin-like proteins has grown considerably, in particular by the discovery of the eukaryotic "actin-related proteins" (ARPs) (Schafer and Schroer 1999) that display an overall 3D structural similarity to actin (Oma and Harata 2011; Robinson et al. 2001), bind and hydrolyze ATP, and show high to moderate sequence similarity to muscle actin (Muller et al. 2005). Not all of the ARPs are expressed in each eukaryote, and depending on their expression and different structural, biochemical and functional properties they can be assigned to discrete subfamilies, identified in amoebae (Machesky et al. 1994), yeast (Poch and Winsor 1997), insects (Fyrberg et al. 1994), vertebrates (Machesky et al. 1997) and plants (Kandassamy et al. 2005). Localisation and presumed biological functions of most of the pro- and eukaryotic actin-like proteins are compiled in Table 1, but it is predicted that still more members of this super-family will be discovered in pro- and eukaryotes. Thus, the cytoskeletal element MreB/ParM/actin is present in pro- and eukaryotes and is an early invention in nature.

Figure 3 shows a phylogenetic tree of representative actin homologues as known of today. Actin and ARPs from yeast to human, MreB from all major bacterial phyla, some ParM and AlfA proteins from bacterial plasmids (both proteins act as filaments in plasmid segregation-machineries (Becker et al. 2006; Moller-Jensen et al. 2003)) and three actin orthologues from archaea were aligned. It has become clear that there are at least three distinct classes of actin-like proteins found in eubacteria (MreB, ParM and AlfA) and one in archaea, and two in eukaryotes, actin and ARPs. Over all sequence homology is poor (only about 15% identity between MreB and muscle actin), as seen from the huge distances between proteins in Fig. 3. ParM, AlfA and MreB show low homology, whereas AlfA, ParM and actin are more closely related. Archaeal actin-like proteins (TA0583, BAB59782 and COG1077) form a cluster that lies in between MreBs and actins. Notably, many archaea do not have a recognizable actin homologue, and many, but not all rod shaped bacteria have MreB, while round species usually do not. On the other hand, actin is found in all eukaryotic genomes we have searched and represents a tight cluster, while ARPs display more diverged sequences. Still, ARPs are highly conserved between lower and higher eukaryotes, and show tighter clustering than bacterial MreB proteins.

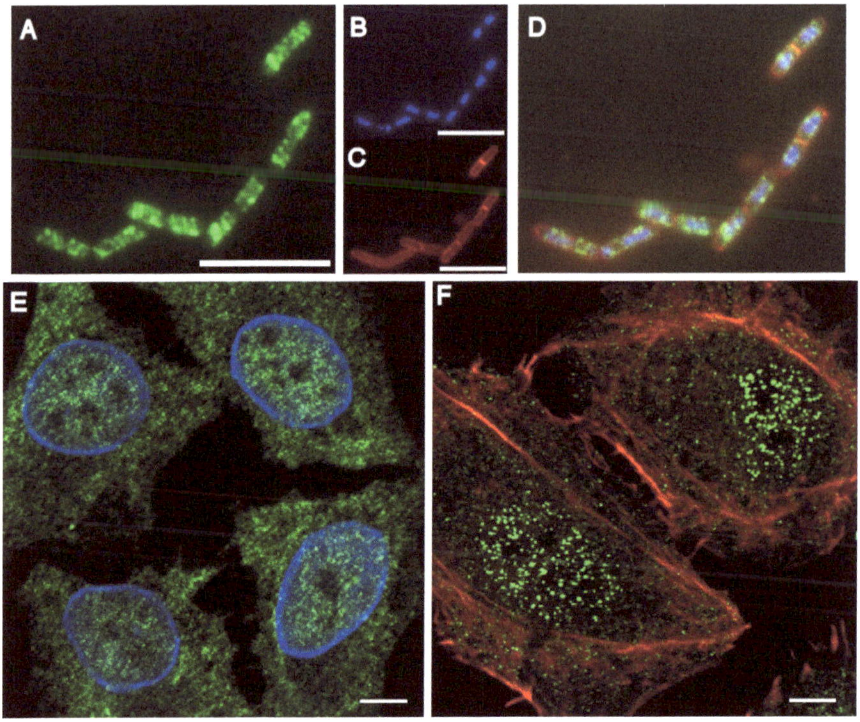

Fig. 2 *Localisation of actin in pro- and eukaryotes*. (**a**–**d**): GFP-MreB expressed from the original gene locus in the model bacterium *Bacillus subtilis*. (**a**): GFP-MreB, (**b**): DNA, (**c**): Membranes, (**d**): Overlay GFP-MreB, DNA and membranes. (**e**–**f**): Actin filaments (F-actin) and other actin assemblies are both seen in HeLa cells (**e**): Confocal sections of cells stained with a monoclonal antibody specific for an actin assembly different from F-actin ("non-F-actin", (Gonsior et al. 1999), see text) show that this actin conformation is abundant in the cytoplasm as well as in the nucleus (*green*). For orientation, the nuclear compartment is revealed by counterstaining with an antibody specific for a protein of the nuclear membrane (*blue*). (**f**): F-actin, as revealed by staining with the toxin phalloidin (*red*), and "non-F-actin" (*green*) are interspersed in the cytoplasm, but only the latter is seen in the nucleus. F-actin is primarily arranged in webs and in large bundles outlining the cell periphery, and forms arrowhead-like structures at cellular attachment sites (courtesy of Cora-Ann Schoenenberger, Maurice Mueller Institute for Structural Biology, University of Basel, Switzerland). White bars: 5 μm

MreB orthologues of distantly related bacteria (*Chlorobium* is more diverged from *Planctomyces* than an elephant from a plant) show 50% sequence identity or more, emphasizing that even MreBs are a highly conserved class of proteins in comparison with other (non-actin related) proteins.

The presence of actins in all kinds of cells and in several eukaryotic cell compartments raises the question of how all these molecules evolved and which pathway the ancestral actin took to invade different cell compartments. By comparing conserved and diverged properties and functions of actin-like proteins, we discuss different scenarios of how actin went on its highly successful journey through evolution.

Table 1 Prokaryotic and eukaryotic actin homologues: location and main physiological activities (for references, see text)

Actin homologues	Organism	Molecular properties	Activities
Bacteria			
FtsA	*E. coli*	Actin-like protein	Cooperates with the tubulin-like FtsZ in cell division
MreB (and paralogues Mbl and MreBH	Rod-shaped bacteria and bacteria with complex cell shape like *B. subtilis, E. coli, Thermotoga maritima, Caulobacter crescentus*	ATP/GTPases, form straight filaments	Control of cell shape and cell polarity, disputed involvement in chromosome segregation
ParM	Plasmid encoded (Gram-negative bacteria)	ATPase that forms F-actin like structures in a right-handed helical twist	Plasmid segregation
AlfA	Plasmid encoded (Gram-positive bacteria)	Forms F-actin-like filaments	Plasmid segregation
MamK	Magnetotactic bacteria	Forms F-actin-like filaments	Magnetosome positioning
Archaea			
Ta0583	*Thermoplasmatales*	ATPase that forms actin-like and MreB-like filaments, but also crystalline sheets	Involved in establishing cell morphology?
Eukaryotes			
Actins	Ubiquitously expressed in all eukaryotic organisms, several isoforms exist as products of discrete genes	ATPases with hexokinase folds, many different ligands, form double helical, twisted filaments, paracrystalline sheets	*Cytoplasmic*: contraction, tension, intracellular transport, locomotion, cell division *Nuclear*: transcription, chromatin remodelling, nucleo-cytoplasmic transport
ARP 1	Humans to yeast	Actin-like ATPase, highest sequence homology to actin, forms filaments	Cytoplasmic, component of the dynactin complex of dyneins

(continued)

Table 1 (continued)

Actin homologues	Organism	Molecular properties	Activities
ARP 2	Humans to plants and fungi	Actin-like ATPase, forms complexes with ARP3	Cytoplasmic, regulates actin polymerisation in complex with ARP3
ARP 3	Humans to plants and fungi	Actin-like ATPase, forms complexes with ARP2	Cytoplasmic, regulates actin polymerisation in complex with ARP2
ARP 4	Human to plants and fungi	Actin-like ATPase, colocalises with nuclear actin	Nuclear, involved in different chromatin remodelling and histone acetyltransferase complexes
ARP 5	Humans to plants and fungi	Actin-like ATPase, several variants	Nuclear, component of different chromatin remodelling and histone acetyltransferase complexes
ARP 6	Yeast, invertebrates, humans and plants	Actin-like ATPase, several variants	Nuclear, component of different chromatin remodelling complexes, heterochromatin and centromeres
ARP 7	Yeast	Yeast-specific actin-like ATPase, forms complexes with actin	Nuclear, involved in chromatin remodelling
ARP 8	Humans to plants, fungi	Actin-like ATPase, several variants	Nuclear, associates with mitotic chromosomes
ARP 9	Yeast	Yeast-specific ARP4-like ATPase, forms complexes with actin	Nuclear, involved in chromatin remodelling
ARP 10	Humans to fungi	Actin-like ATPase, several variants	Cytoplasmic, involved in dynein motor activity
ARP 11			Cytoplasmic, component of the dynactin complex of dyneins

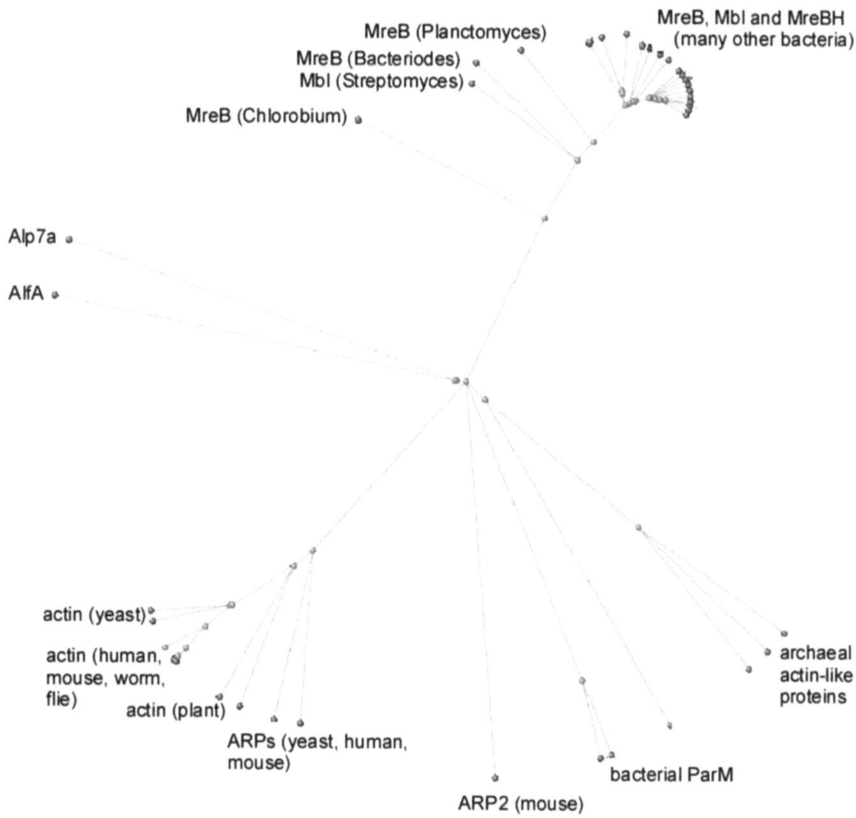

Fig. 3 Phylogenetic tree of selected actin-like proteins from all three domains of life. The following sequences were aligned using COBALT (http://www.ncbi.nlm.nih.gov/tools/cobalt/cobalt.cgi?link_loc=BlastHomeLink): Eukarya: actin from *Saccharomyces cerevisiae*, *Schizosaccharomyces pombe*, *Mus musculus*, *Caenorhabditis elegans*, *Arabidopsis thaliana*, actinActA1 [*Homo sapiens*], ACT-2 [*Caenorhabditis elegans*], actin5C, isoform_C [*Drosophila melanogaster*], ARP1 [*Homo sapiens*], ARP2 [*Saccharomyces cerevisiae*], ARP3 [*Homo sapiens*], ARP3 [*Mus musculus*], ARP2[*Mus musculus*], Bacteria: ParM [*Escherichia coli*], ParM [*Staphylococcus aureus*], ParM [*Klebsiella pneumoniae*], MreB [*Caulobacter crescentus*], MreB [*Thiobacillus denitrificans*], MreB [*Escherichia coli* BL21], MreB [*Pseudomonas aeruginosa*], MreB [*Stigmatella aurantiaca*], MreB [*Campylobacter lari*], MreB [*Helicobacter pylori*], MreB [*Lactobacillus acidophilus*], MreB [*Bacillus subtilis*], MreB [*Clostridium tetani*], MreB [*Streptomyces griseus*], MreB [*Synechococcus elongatus*], MreB [*Chlamydia trachomatis*], MreB [*Planctomyces maris*], MreB [*Bacteroides fragilis*], MreB [*Chlorobium tepidum*], MreB [*Chloroflexus aurantiacus*], MreB [*Thermotoga maritima*], MreB [*Thermotoga petrophila*], MreB [*Aquifex aeolicus*], Mbl [*Bacillus subtilis*], MreBH [*Bacillus subtilis*], Mbl [*Streptomyces coelicolour*], Alp7A [*Bacillus subtilis*], AlfA [*Bacillus subtilis*]. Archaea: TA0583 [*Thermoplasma acidophilum*], COG1077 [*Ferroplasma acidarmanus*], BAB59782 [*Thermoplasma volcanium*]. Standard settings were used for alignment (e.g. gap penalties −11, −1) and for tree construction (fast minimum evolution method, maximal sequence difference 0.9), whole protein sequences were used

2 Actin Filament-Based Dynamics in Eukaryotic and Prokaryotic Cells

As already mentioned, eukaryotic actins play a key role in cellular motility. They are vital for intracellular processes such as vesicle generation and transport, shape changes, cell division, contraction, adhesion and, in case of motile cells, also for locomotion. These events require a time- and space-controlled regulation of actin filament generation and subsequent function in various regions of the cell. To achieve this, actins and many of the ARPs can interact with a vast catalogue of ligand proteins. Inhibition and stimulation of filament formation, bundling and network formation, as well as membrane attachment in different parts of the cell are regulated by such complexes (Chhabra and Higgs 2007; Pollard and Cooper 1986). Subsequent motility is the consequence of two different processes: (1). It can be achieved by rapid polymerization of ATP-charged actin to the conventional, double helical F-actin that, when localized in distal parts of the cytoplasm, pushes the plasma membrane forward. This is for example the basis for rapid changes of cell shape, as seen during generation of cellular filopodia or dendrites of neuronal cells, and in pollen germination in plant cells. (2). Cell polarity, cell division, adhesion and cell–cell contacts require the interaction of the unipolar F-actin filaments with bipolar myosin filaments. Actomyosin interaction is the basis for muscle contraction and is indispensable for wound healing and tissue formation.

The discovery that prokaryotic actins are able to form polar filaments with striking structural resemblance to F-actin has prompted many studies on their function in spatially organised processes, like cell shape maintenance and cell cycle progression. MreB, the best characterized prokaryotic actin-like protein, is essential in many rod shaped bacteria and bacteria with a more complex cell shape (curved/vibrio shaped). The depletion of MreB in *Bacillus subtilis* or *Escherichia coli* (model organisms for Gram positive or Gram negative bacteria, respectively) leads to the formation of round cells that continue to grow for some time and reinitiate DNA replication until lysis occurs (Graumann 2007). However, so far, no interaction of prokaryotic filaments with motor proteins has been found (see below) and thus, the role of MreB and its relatives in regulating cell morphology and possibly chromosome segregation is assumed to be the sole consequence of formation of filaments that direct the localization of cell-wall synthesizing as well as of cytosolic enzymes. Indeed, incorporation of new cell wall material appears to occur in a helical pattern (Daniel and Errington 2003), and MreB interacts with membrane proteins involved in cell shape maintenance and in cell wall synthesis (Defeu Soufo and Graumann 2006; Kawai et al. 2009; Kruse et al. 2005) that may be positioned in a helical arrangement through MreB filaments.

Furthermore, in eukaryotes, regulation of actin based motility requires an additional process so far not seen in prokaryotes: the filaments and their supramolecular assemblies are linked to different signalling cascades. Protein members of these cascades are activated by small GTPases and harbour specific domains that regulate actin filament assembly and dynamics, they act as "nucleators" (Campellone and Welch 2010; Chhabra and Higgs 2007). Some of the ARPs can also control F-actin filament assembly (Table 1). In this way, eukaryotic cells are able to respond to their environment by rapid local rearrangement and fine tuning of their actin filament complement.

3 Involvement of Eukaryotic Actins in Nuclear Functions

Eukaryotic actin is a dual compartment protein, residing in the cytoplasm as well as in the nucleus of protozoa, animals and plants (reviewed in (Jockusch et al. 2006)). Notably, some of the ARPs were found exclusively in the nucleus (Oma and Harata 2011). Nuclear functions of actin and several ARPs include activation and processing of genes for transcription: chromatin alignment and remodelling, optimising the activity of RNA polymerases and transport of their initial products. Consistent with a prominent role in euchromatin conditioning for transcription, actin was localized at decondensed chromatin and in the nucleolus of mammalian nuclei (Dingova et al. 2009). Biochemical analyses consistently identified actin and specific ARPs as components of the ATPase and acetyl transferase complexes that contribute to the gigantic protein assemblies involved in chromatin remodelling from yeast to man (Blessing et al. 2004; Chen and Shen 2007; Dion et al. 2010; Olave et al. 2002), but their precise role in this location is not known. All three eukaryotic RNA polymerases, i.e. the enzyme complexes executing transcription of ribosomal RNA, mRNA and t-RNAs contain actin (Grummt 2006; Percipalle and Visa 2006). The most detailed studies deal with RNA polymerase II, and here, actin has been found to regulate the activity of this enzyme during mRNA transcription in fission yeast (Mitsuzawa et al. 2005), amphibian oocytes (Scheer et al. 1984) and in mammalian cells (Hofmann et al. 2004). The association of actin with RNA polymerases may precede, be simultaneous with or follow actin-associated chromatin remodelling (Grummt 2006; Percipalle and Visa 2006).

In addition to actin, many of the actin-binding proteins that interact with cytoplasmic actin and regulate assembly and dynamics of the conventional F-actin filaments are also found in the nucleus, including some of the nucleators (Gieni and Hendzel 2009). This may suggest that nuclear actin is filamentous and can be arranged into various supramolecular, dynamic structures, as required for motility, analogous to the cytoplasmic assemblies. However, so far, there is no solid evidence to support this concept. First of all, it is not known whether actin in the nucleus is monomeric, filamentous as in F-actin, or present in another supramolecular form yet to be defined. Notably, actin antibodies that do not recognize native F-actin but rather ill-defined “exotic” actin conformations identify actin faithfully within the euchromatin region of mammalian nuclei ((Gonsior et al. 1999; Schoenenberger et al. 2005) and Fig. 2). Second, although a specific nuclear myosin has been identified (Pestic-Dragovich et al. 2000) that is also associated with chromatin-remodelling complexes and with RNA polymerases (Fomproix and Percipalle 2004; Grummt 2006; Zhao et al. 2008), this myosin belongs to the class of small myosins that are unable to form the bipolar filaments required for actomyosin-based contractility.

Therefore, so far, any assumption on actomyosin contraction participating in chromatin rearrangement or transcription must remain purely speculative (Gieni and Hendzel 2009; Hofman and de Lanerolle 2006).

In addition to the specific nuclear myosin I, there are also a few actin binding proteins exclusively found in the nucleus (Ankenbauer et al. 1989; Rimm and Pollard 1989), but virtually nothing is known about their function.

4 Comparing Pro- and Eukaryotic Actins: Dissimilarities and Similarities

A clear distinction between MreB and actin is seen in their biochemical properties: the formation of F-actin filaments under physiological conditions strictly depends on ATP and magesium, while MreB from two bacterial species can also use GTP to form filaments (albeit GTP is less efficient than ATP). Furthermore, the influence of monovalent and divalent cations, temperature and pH on filament assembly differs between pro- and eukaryotic actins, but also between different prokaryotic actin homologues (Mayer and Amann 2009). Spontaneous nucleation of MreB occurs at 25- to 100-fold lower protein concentration than that required for actin (Bean and Amann 2008; Esue et al. 2005). Thus, while eukaryotic actin filament formation is regulated by nucleating proteins, prokaryotic actins like MreB may assemble in bacterial cells due to protein or ATP levels. However, the existence of bacterial nucleators cannot be excluded. In any event, available biochemical data on MreB suggest that it has more "ancient" properties for the formation of dynamic filaments, as the bacterial cell may have less complex requirements for the action of MreB than the eukaryote has for actin. Interestingly, MreB has recently been shown to also confer mechanical rigidity to bacterial cells (Defeu Soufo and Graumann 2010; Wang et al. 2010), another parallel to eukaryotic actin, even though MreB does not form a continuous helix but rather many short helical filaments showing high turnover (Defeu Soufo and Graumann 2005).

Different from F-actin, MreB and ParM filaments can grow and shrink at both ends, and show catastrophic decay very similar to microtubules (Defeu Soufo and Graumann 2010; Garner et al. 2004). ParM is encoded on many low copy number plasmids that require an active partitioning system for their propagation into the daughter cells. Through the addition of ParM monomers at the filament tips, and through an interaction with an adaptor protein binding to the partitioning site on the plasmid, duplicated plasmids are moved away from the cell centre to opposite cell poles in a simple mitotic-like machinery (Gamer et al. 2007; Garner et al. 2004; Moller-Jensen et al. 2003; Salje et al. 2009). The expansive abilities of ParM reveal how huge the repertoire of polymerization competence of actin-like proteins can be, and how far this protein family has diverged.

Likewise, an important difference between pro- and eukaryotic actin filaments concerns the fact that the latter use the interaction with motor proteins to achieve various aspects of motility. None of the classical motor proteins found in eukaryotic cells, like the actin-associated myosins, or the microtubule-associated kinesins or dyneins seem to be present in bacterial cells (although bacteria contain proteins related to the eukaryotic dynamins, large multidomain GTPases cooperating with actin in vesicle formation (Low and Loewe 2006)).

One protein that interacts with both, pro-and eukaryotic actins, has already been analysed in greater detail. MreB and actin both bind the translation elongation factor 1α (EF-Tu in bacteria) in vivo and in vitro (Defeu Soufo and Graumann 2010; Gross and Kinzy 2005; Liu et al. 1996). EF-Tu has long been proposed to also play a role as a cytoskeletal element in bacteria. In vitro, it spontaneously forms filaments, and it has been shown to be associated with the bacterial cell membrane (Beck et al. 1978; Jacobson and Rosenbusch 1976). In *B. subtilis* cells, MreB and EF-Tu interact along the bundles of protofilaments running underneath the membrane in a helical pattern. In addition, EF-Tu is also present at subcellular spaces close to the cell poles, where most of the ribosomes are located. Indirect evidence suggests that EF-Tu is required for the proper localization of MreB in filament bundles, while MreB is clearly required to position EF-Tu at the membrane. It has been proposed that EF-Tu filaments may serve as tracks for the remodelling of MreB filaments (Defeu Soufo and Graumann 2010). In the eukaryotic setting, EF 1α also interacts with actin along F-actin filaments, and in vitro, EF 1α acts as an F-actin bundling factor (Murray ct al. 1996).

Quite analogous to eukaryotic actins being associated with RNA polymerases (see above), MreB has also been shown to interact with RNA polymerase (RNAP) in *E. coli* (Kruse et al. 2006), as well as with topoisomerase IV (Madhabhushi and Marains 2009). There is an ongoing debate whether MreB is involved in chromosome segregation, with as many positive as negative pieces of evidence having been put forward. It is possible that a loss of function of MreB in cell wall maintenance may directly or indirectly affect chromosome segregation, as discussed in (Graumann 2009). Many more experiments are required to shed light on the MreB/RNAP/chromosome segregation connection, but the relation of actin and MreB to the transcription machinery remains intriguing.

5 Evolution of the Eukaryotic Cell

16S RNA analysis has suggested that all cells derive from one last universal common ancestor, called LUCA, that must have been able to self propagate and survive in the then very harsh environment (Fig. 4). LUCA most likely fed on reduced inorganic compounds, i.e. was chemolithotroph, because recent evidence suggests that growth on organic compounds evolved later than lithotrophic metabolism (oxidation of inorganic compounds, (Say and Fuchs 2010)). LUCA then evolved into two very different cell types, bacteria and archaea, the most striking difference being completely different membrane systems and cell wall structures, amongst many other differences. Phylogenetically, eukaryotes then evolved from archaea, but how the eukarya arose is still under intensive debate. Many theories have been put forward, all of

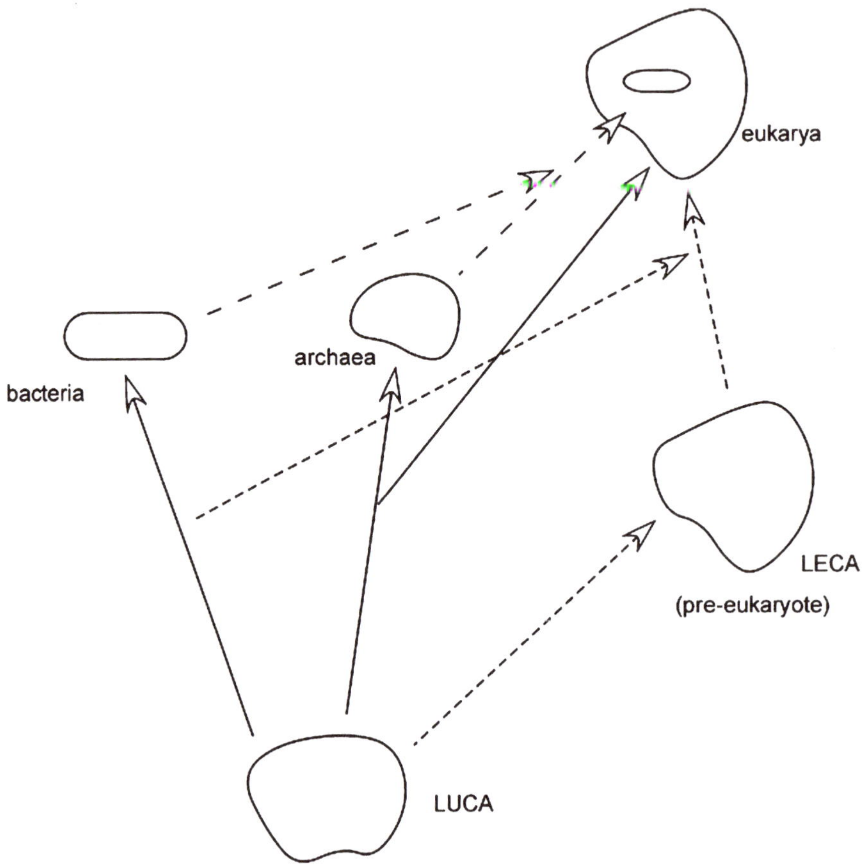

Fig. 4 Model for the evolution of eukaryotic cells. LUCA = last universal common ancestor, LECA = last eukaryote common ancestor. Solid lines indicate the phylogenetic tree according to 16S rRNA analyses, loosely dashed lines the theory of fusion of an archaeon with a bacterium, or the predation of an archaeon by a bacterium; tightly dashed lines indicate the evolution of a pre-eukaryote, which was then predated by a bacterium that through symbiosis then evolved into the mitochondrion (indicated by the rod within the eukaryotic cell)

which have their charms but also their shortcomings. A set of popular theories is based on the idea that an archaeal cell fused with a bacterial cell, retaining the archaeal transcription and translation machinery, and the bacterial metabolism and cell membrane (Fig. 4, loosely dashed lines). However, both cell types have rather rigid cell walls (with the exception of the wall less bacteria of the *Mollicutes*), so cell fusion is difficult to envision. A newer hypothesis proposes predation and ensuing symbiosis as the starting point for eukaryote evolution (Davidov and Jurkevitch 2009). The mitochondrion is most closely related to the group of "alpha" proteobacteria, a large branch of bacteria that contains many predatory organisms. Some of these squeeze themselves through the cell wall and thrive in the periplasm, from which they feed on the host cell. Such a predatory bacterium (performing oxidative phosphorylation) may have infected an archaeon, to become converted into a symbiont that ultimately evolved into the mitochondrion, giving away most of its DNA to be joined with the archaeal genome. However, this theory assumes that the archael isoprenoid-ether based membranes became converted into the bacterial lipid esters, which is very hard to envision. In parallel, all membrane proteins, adapted to the much more rigid and sealed archaeal membrane, would have had to evolve into functions in a very different hydrophobic environment, a scenario that is hardly feasible.

Another set of arguments must be considered. Many studies have identified a complement of genes common to all (or at least almost all) eukaryotic cells, which lack homologues in prokaryotic cells. These genes must have been present in a last eukaryote-common ancestor (LECA, (Margulis et al. 2006)), as shown in Fig. 4. Amongst these are genes for kinesins, myosins, clathrin, various signalling proteins and many others (Hartman and Fedorov 2002). These assumptions and many additional data argue for a third line of organisms ("chronocytes" or pre-eukaryotes) that evolved – at least for a while – in parallel with eubacteria and archaea, and that developed an already sophisticated intracellular motor machinery as well as proteins mediating endocytosis, protein degradation, and different signaling systems (Hartman and Fedorov 2002). These cells must have had bacterial-like membranes and anaerobic metabolism (archaea have greatly diverged glycolytic or other types of metabolic pathways), but an archaeal-type transcription machinery. We propose that LECA was attacked by an alpha-proteobacterial parasite, which after invasion failed to kill the host, but developed into an endosymbiont that provided oxidative phosphorylation and later developed into the mitochondrion, marking the generation of the first modern eukaryotic cells. While these flourished and evolved, the original pre-eukaryote with its slower anaerobic metabolism died out.

6 The Pathway of the Actin Ancestor

Based on the fact that actin-like proteins are present in bacteria, archaea and eukarya, and that in all cases, the orthologues form dynamic filamentous structures that in bacteria and eukaryotes conduct dynamic as well as structural functions, strongly suggests that actin was present in LUCA and performed an important or essential task. From LUCA, actin evolved into the three domains of life (Fig. 4). Looking at all diverse functions performed by actin like proteins (Table 1), one realizes that the evolutionary development of actin-like proteins is a story of great success. While MreB and ParM could greatly diverge in bacteria; actin itself remained highly conserved in eukaryotes (Fig. 3), but ARPs evolved in eukaryotes (and apparently not in prokaryotes) to take over many additional functions.

Actin may already have been necessary for cytokinesis in the pre-eukaryote, while there is no evidence for actin-like proteins being involved in cytokinesis of modern prokaryotes. Interestingly, one of the two major branches of archaea use an ESCRT like mechanism (ESCRT pinches off vesicles at the Golgi apparatus) for cell division (Samson et al. 2008), rather than the FtsZ (tubulin)-based mechanism of most other prokaryotes (Bi and Lutkenhaus 1991), underlining the enormous versatility of cytoskeletal elements and of filament-forming proteins to perform fundamental tasks in different ways.

Since the actin filament system is able to perform vital motility processes by simple polymerization without motor proteins, and because bacteria lack known motor proteins (see above), myosins likely appeared later in evolution than actin, and not in parallel (Mitchison 1995). Interestingly, large scale genomic sequencing revealed that genes for actin isoforms, regulators of actin filaments and supramolecular assemblies, ARPs, myosins and nucleators like formins are universal throughout all existing eukaryotic species (Cvrckova et al. 2004; Rivero and Cvrckova 2007). Hence, one can conclude that the full complement of the actin cytoskeleton is an early invention in the eukaryotic lineage and was present in LECA.

The nucleus is thought to have evolved through membrane invagination around the genome. As MreB is also associated with RNAP and topoisomerase in modern bacteria, and ParM is the driving force in plasmid segregation, it is reasonable to assume that actin has also carried out vital tasks in transcription in LUCA. After the formation of the nucleus, actin must have been able to shuttle in and out of the nucleus. Indeed, actin import and export systems that can control diffusion have been described (reviewed in (Ambrosino et al. 2010; Jockusch et al. 2006; Skarp and Vartiainen 2010)). The fact that also some of the ARPs are able to shuttle while others are strictly nuclear (Table 1) supports the idea that nuclear actin is not a product of recent evolution, but an ancient trait of eukaryotic cells that co-evolved with the establishment of the nucleus.

As true actins show little sequence divergence between yeast, worm and mammals (Fig. 3), it is plausible that the actin sequence has been under considerable restraint during evolution, while ARPs with their much higher sequence diversity (Fig. 3) have been less so. ARPs usually have a more specialized function than actin, and can therefore evolve with average speed. In contrast, the enormous diversity of actin's functions, as described above, puts a restraint on its pace of evolution. We favour the idea that an additional restraint for actin evolution has been its ancient function in the nucleus. Given that special nuclear ARPs had the time to evolve, but no specialized nuclear actin, suggests that actin has had an ancient role on the DNA (Skarp and Vartiainen 2010) that was simply retained after the "invention" of the nucleus.

However, cytoskeletal proteins show an amazing functional plasticity: tubulin forms the spindle apparatus in eukaryotic cells, while its orthologue, FtsZ, forms the initial cytokinetic ring in most bacteria and in one of the two major branches of the archaea. Recently, an FtsZ/tubulin protein has been shown to push plasmids into opposite cell halves in a bacterium (Larsen et al. 2007), very similar to ParM and AlfA. On the other hand, actin forms the cleavage ring in dividing cells of fungi, amoebae and animals, whereas MreB is not directly involved in cell division but in maintenance of cell shape, and possibly in chromosome segregation in some bacteria. Thus, in this case, the functions of tubulins and actins have been swapped between pro- and eukaryotes. Among the bacterial actin-like proteins, we also observe an amazing functional versatility: bacterial ParM actively pushes plasmids away from the cell centre, interacting with a centromere-binding protein that regulates ParM

ATPase activity. In many rod shaped bacteria, MreB is a key player in cell shape maintenance (interacting with many membrane proteins), whereas in *Helicobacter pylori*, the loss of MreB does not affect helical cell shape, but the progression through the cell cycle and – directly or indirectly – the activity of an enzyme that is vital for the infectivity of the human pathogen (Waidner et al. 2009). These cases illustrate that actin-like proteins can adopt new functions and give up others, and that therefore, the argument about the appearance of nuclear actin during evolution could be very different: the acquisition of actin in the nucleus (and its interaction with RNAP and topoisomerase in some bacteria) could be a rather recent development during evolution, a fine tuning rather than a fundamental function in transcription, thus explaining why nuclear actin has not diverged from cytosolic actin. However, the many pieces of evidence outlined above argue for the concept that the interaction of actin-like proteins with chromosomes is indeed ancient.

7 Conclusion

Actin-like proteins are evolutionary ancient and must have been present in the last common ancestor of all cells, where with little doubt they performed vital tasks based on the ability to form dynamic filaments that can confer dynamic as well as structural tasks. Actin-like proteins have greatly diverged to perform an amazing array of functions, many of which are different between pro- and eukaryotes, but several of which are conserved. Based on current knowledge, we propose that actin plays a conserved and ancient function in the nucleus that we speculate has evolved long after actin was acting on the DNA. Understanding the mode of action of actin-like proteins in much more detail, their regulation and their interacting partners will tell us a lot about evolution. Therefore, it will be highly interesting to further study actin, MreB and ParM activities and functions, as these will give us not only a deeper comprehension of molecular and cellular biology, but also more and more clues about the evolution of cells and of cellular compartments.

Acknowledgments We thank Andreas Bracher (Munich, Germany) for Fig. 1, Joel Defeu Soufo (Freiburg, Germany) for Fig. 2a–d and Cora-Ann Schoenenberger (Basel) for Fig. 2e–f. Our own research mentioned in this article was funded by the German Research Council (DFG) and the Fonds der Chemischen Industrie. Due to space limitations, the Reference List comprises only selected publications and we apologize to the many authors whose work could not be cited.

References

Ambrosino C, Tarallo R, Bamundo A, Cuomo D, Franci G, Nassa G, Paris O, Ravo M, Giovane A, Zambrano N, Lepikhova T, Jänne OA, Baumann M, Nyman TA, Cicatello L, Weisz A (2010) Identification of a hormone-regulated dynamic nuclear actin network associated with estrogen receptor alpha in human breast cancer cell nuclei. Mol Cell Proteomics 9:1352–1367

Ankenbauer T, Kleinschmidt JA, Walsh MJ, Weiner OH, Franke WW (1989) Identification of a widespread nuclear actin binding protein. Nature 342:822–825

Bean GJ, Amann KJ (2008) Polymerization properties of the *Thermotoga maritima* actin MreB: roles of temperature, nucleotides, and ions. Biochemistry 47:826–835

Beck BD, Arscott PG, Jacobson A (1978) Novel properties of bacterial elongation factor Tu. Proc Natl Acad Sci USA 75:1250–1254

Becker E, Herrera NC, Gunderson FQ, Derman AI, Dance AL, Sims J, Larsen RA, Pogliano J (2006) DNA segregation by the bacterial actin AlfA during Bacillus subtilis growth and development. Embo J 25:5919–5931

Bi EF, Lutkenhaus J (1991) FtsZ ring structure associated with division in Escherichia coli. Nature 354:161–164

Blessing CA, Ugrinova GT, Goodson HV (2004) Actin and ARPs: action in the nucleus. Trends Cell Biol 14:435–442

Bork P, Sander C, Valencia A (1992) An ATPase domain common to prokaryotic cell cycle proteins, sugar kinases, actin and hsp70 heat shock proteins. Proc Natl Acad Sci USA 89:7290–7294

Campellone KG, Welch MD (2010) A nucleator arms race: cellular control of actin assembly. Nat Rev Mol Cell Biol 11:237–251

Carballido-Lopez R, Errington J (2003) The bacterial cytoskeleton: *in vivo* dynamics of the actin-like protein Mbl of *Bacillus subtilis*. Dev Cell 4:19–28

Chen M, Shen X (2007) Nuclear actin and actin-related proteins in chromatin dynamics. Curr Opin Cell Biol 19:326–330

Chhabra ES, Higgs HN (2007) The many faces of actin: matching assembly factors with cellular structures. Nat Cell Biol 9:1110–1121

Cvrckova F, Rivero F, Bavinka B (2004) Evolutionarily conserved modules in actin nucleation: lessons from *Dictyostelium discoideum* and plants. Review article. Protoplasma 224:15–31

Daniel RA, Errington J (2003) Control of cell morphogenesis in bacteria: two distinct ways to make a rod-shaped cell. Cell 113:767–776

Davidov Y, Jurkevitch E (2009) Predation between prokaryotes and the origin of eukaryotes. BioEssays 31:748–757

Defeu Soufo HJ, Graumann PL (2005) Bacillus subtilis actin-like protein MreB influences the positioning of the replication machinery and requires membrane proteins MreC/D and other actin-like proteins for proper localization. BMC Cell Biol 6:10

Defeu Soufo HJ, Graumann PL (2006) Dynamic localization and interaction with other Bacillus subtilis actin-like proteins are important for the function of MreB. Mol Microbiol 62:1340–1356

Defeu Soufo HJ, Graumann PL (2010) *Bacillus subtilis* MrB paralogues have different filament architectures and lead to shape remodelling of a heterogogous cell system. Mol Microbiol 78:1145–1158

Defeu Soufou HJ, Graumann PL (2004) Dynamic movement of actin-like proteins within bacterial cells. EMBO Rep 5:789–794

Dingova H, Fukalova J, Maninova M, Philimonenko VV, Hozak P (2009) Ultrastructural localization of actin and actin-binding proteins in the nucleus. Histochem Cell Biol 131:425–434

Dion V, Shimada K, Gasser SM (2010) Actin-related proteins in the nucleus: life beyond chromatin remodelers. Curr Opin Cell Biol 22:383–391

Esue O, Cordero M, Wirtz D, Tseng Y (2005) The assembly of MreB, a prokaryotic homolog of actin. J Biol Chem 280:2628–2635

Figge RM, Divakaruni AV, Gober JW (2004) MreB, the cell shape-determining bacterial actin homologue, co-ordinates cell wall morphogenesis in *Caulobacter crescentus*. Mol Microbiol 51:1321–1332

Fomproix N, Percipalle P (2004) An actin-myosin complex on actively transcribing genes. Exp Cell Res 294:140–148

Fyrberg C, Ryan L, Kenton M, Fyrberg E (1994) Genes encoding actin-related proteins of *Drosophila melanogaster*. J Mol Biol 241:498–503

Gamer EC, Campbell CS, Weiberl DB, Mullins RD (2007) Reconstitution of DNA segregation driven by assembly of a prokaryotic actin homolog. Science 315:1270–1274
Garner EC, Campbell CS, Mullins RD (2004) Dynamic instability in a DNA-segregating prokaryotic actin homolog. Science 306:1021–1025
Gieni RS, Hendzel MJ (2009) Actin dynamics and functions in the interphase nucleus: moving toward an understanding of nuclear polymeric actin. Biochem Cell Biol 87:283–306
Gonsior SM, Platz S, Buchmeier S, Scheer U, Jockusch BM, Hinssen H (1999) Conformational differences between nuclear and cytoplasmic actin as detected by a monoclonal antibody. J Cell Sci 112:797–809
Graumann PL (2007) Cytoskeletal elements in bacteria. Annu Rev Microbiol 61:589–618
Graumann PL (2009) Dynamics of bacterial cytoskeletal elements. Cell Motil Cytoskeleton 66:909–914
Gross SR, Kinzy TG (2005) Translation elongation factor 1A is essential for regulation of the actin cytoskeleton and cell morphology. Nat Struct Mol Biol 12:772–778
Grummt I (2006) Actin and myosin as transcription factors. Curr Opin Genet Dev 16:191–196
Hartman H, Fedorov A (2002) The origin of the eukaryotic cell: a genomic investigation. Proc Natl Acad Sci USA 99:1420–1425
Hofman WA, de Lanerolle P (2006) Nuclear actin: to polymerize or not to polymerize. J Cell Biol 172:495–496
Hofmann WA, Stojiljkovic L, Fuchsova B, Vargas GM, Mavrommatis E, Philimonenko V, Kysela K, Goodrich JA, Lessard JL, Hope TJ, Hozak P, de Lanerolle P (2004) Actin is part of pre-initiation complexes and is necessary for transcription by RNA polymerase II. Nat Cell Biol 6:1094–1101
Jacobson GR, Rosenbusch JP (1976) Abundance and membrane-association of elongation factor Tu in *E. coli*. Nature 261:23–26
Jockusch BM, Schoenenberger CA, Stetefeld J, Aebi U (2006) Tracking down the different forms of nuclear actin. Trends Cell Biol 16:391–396
Jones LJ, Carbadillido-Lopez R, Errington J (2001) Control of cell shape in bacteria: helical, actin-like filaments in *Bacillus subtilis*. Cell 104:913–922
Kabsch W, Mannherz HG, Suck D, Pai EF, Holmes KC (1990) Atomic structure of the actin:DNase I complex. Nature 347:37–44
Kandassamy MK, McKinney EC, Deal RB, Meagher RB (2005) Arabidopsis ARP7 is an essential actin-related protein required for normal embryogenesis, plant architecture, and floral organ abscission. Plant Physiol 138:2019–2032
Kawai Y, Asai K, Errington J (2009) Partial functional redundancy of MreB isoforms, MreB, Mbl and MreBH, in cell morphogenesis of Bacillus subtilis. Mol Microbiol 73:719–731
Kim SY, Gittal Z, Kinkhabwala A, Shapiro L, Moerner WE (2006) Single molecules of the bacterial actin MreB undergo directed treadmilling motion in *Caulobacter crescentus*. Proc Natl Acad Sci USA 103:10929–10934
Kruse T, Moeller-Jensen J, Loebner-Olesen A, Gerdes K (2003) Dysfunctional MreB inhibits chromosome segregation in *Escherichia coli*. Embo J 22:5283–52892
Kruse T, Bork-Jensen J, Gerdes K (2005) The morphogenetic MreBCD proteins of Escherichia coli form an essential membrane-bound complex. Mol Microbiol 55:78–89
Kruse T, Blagoev B, Lobner-Olesen A, Wach M, Sasaki K (2006) Actin homolog MrB and RNA polymerase interact and are both required for chromosome segregation in *Escherichia coli*. Genes Dev 20:113–124
Larsen RA, Cusumano C, Fujioka A, Lim-Fong G, Patterson P, Pogliano J (2007) Treadmilling of a prokaryotic tubulin-like protein, TubZ, required for plasmid stability in *Bacillus thuringiensis*. Genes Dev 21:1340–1352
Liu G, Tang J, Edmonds BT, Murray J, Levin S, Condeelis J (1996) Bundling of actin filaments by elongation factor 1 alpha inhibits polymerization at filament ends. J Cell Biol 135:1309–1321
Low HH, Loewe J (2006) A bacterial dynamin-like protein. Nature 444:766–769
Machesky LM, Atkinson SJ, Ampe C, Vandekerckhove J, Pollard TD (1994) Purification of a cortical complex containing two unconventional actins from *Acanthamoeba* by affinity chromatography on profilin-agarose. J Cell Biol 127:107–115

Machesky LM, Reeves E, Wientjes F, Mattheyse FJ, Grogan A, Totty NF, Burlingame AL, Hsuan JJ, Segal AW (1997) Mammalian actin-related protein 2/3 complex localizes to regions of lamellipodial protrusion and is composed of evolutionarily conserved proteins. Biochem J 328:105–112

Madhabhushi R, Marains KJ (2009) Actin homolog MreB affects chromosome segregation by regulating topoisomerase IV in *Escherichia coli*. Mol Cell 33:171–180

Margulis L, Chapman M, Guerrero R, Hall J (2006) The last eukaryotic common ancestor (LECA): acquisition of cytoskeletal motility from aerotolerant spirochetes in the proterozoic Eon. Proc Natl Acad Sci USA 103:13080–13085

Mayer JA, Amann KJ (2009) Assembly properties of the *Bacillus subtilis* actin, MreB. Cell Motil Cytoskeleton 66:109–118

Mitchison TJ (1995) Evolution of a dynamic cytoskeleton. Philos Trans R Soc Lond B Biol Sci 349:299–304

Mitsuzawa H, Kimura M, Kanda E, Ishihama A (2005) Glyceraldehyde-3-phosphate dehydrogenase and actin associate with RNA polymerase II and interact with its Rpb7 subunit. FEBS Lett 579:48–52

Moller-Jensen J, Borch J, Dam M, Jensen RB, Roepstorff P, Gerdes K (2003) Bacterial mitosis: ParM of plasmid R1 moves plasmid DNA by an actin-like insertional polymerization mechanism. Mol Cell 12:1477–1487

Muller J, Oma Y, Vallar L, Friederich E, Poch O, Windsor B (2005) Sequence and comparative genomic analysis of actin-related proteins. Mol Biol Cell 16:5736–5748

Murray JW, Edmonds BT, Liu G, Condeelis J (1996) Bundling of actin filaments by elongation factor 1 alpha inhibits polymerization at filament ends. J Cell Biol 135:1309–1321

Oda T, Iwasa M, Alhara T, Maeda Y, Narita A (2009) The nature of the globular- to fibrous-actin transition. Nature 457:441–445

Olave IA, Reck-Peterson SL, Crabtree GR (2002) Nuclear actin and actin-related proteins in chromatin remodeling. Annu Rev Biochem 71:755–781

Oma Y, Harata M (2011) Actin-related proteins localized in the nucleus: from discovery to novel roles in nuclear organization. Nucleus (in press)

Orlowa A, Garner EC, Galkin VE, Heuser J, Mullins RD, Egelman EH (2007) The structure of bacterial ParM filaments. Nat Struct Mol Biol 14:921–926

Percipalle P, Visa N (2006) Molecular functions of nuclear actin in transcription. J Cell Biol 172:967–971

Pestic-Dragovich L, Stojiljkovic L, Philimonenko AA, Nowak G, Ke Y, Sttlage RE, Shabanowitz J, Hunt DF, Hozak P, de Lanerolle P (2000) A myosin I isoform in the nucleus. Science 290:337–341

Poch O, Winsor B (1997) Who's who among the *Saccharomyces cerevesiae* actin-related proteins? A classification and nomenclature proposal for a large family. Yeast 13:1053–1058

Pollard TD, Cooper JA (1986) Actin and actin-binding proteins. A critical evaluation of mechanisms and functions. Annu Rev Biochem 55:987–1035

Popp D, Narita A, Oda T, Fujisawa T, Matsuo H, Nitanai Y, Iwasa M, Maeda K, Onishi H, Maeda Y (2008) Molecular structure of the ParM polymer and the mechanism leading to its nucleotide-driven dynamic instability. Embo J 27:570–579

Popp D, Narita A, Maeda K, Fujisawa T, Ghoshdastider U, Iwasa M, Maeda Y, Robinson RC (2010) Filament structure, organization and dynamics in MreB sheets. J Biol Chem 285:15858–15865

Rimm DL, Pollard TD (1989) Purification and characterization of an Acanthamoeba nuclear actin-binding protein. J Cell Biol 109:585–591

Rivero F, Cvrckova F (2007) Origins and evolution of the actin cytoskeleton. Adv Exp Med Biol 607:97–110

Robinson RC, Turbedsky K, Kaiser DA, Marchand JB, Higgs HN, Choe S, Pollard TD (2001) Crystal structure of Arp2/3 complex. Science 294:1660–1661

Roeben A, Kofler C, Nagy I, Nickell S, Hartl FU, Bracher A (2006) Crystal structure of an archaeal actin homolog. J Mol Biol 358:145–156

Salje J, Zuber B, Löwe J (2009) Electron cryomicroscopy of *E.coli* reveals filament bundles involved in plasmid DNA segregation. Science 323:509–512

Samson RY, Obita T, Freund SM, Williams RL, Bell SD (2008) A role for the ESCRT system in cell division in archaea. Science 322:1710–1713

Say RF, Fuchs G (2010) Fructose 1,6-bisphosphate aldolase/phosphatase may be an ancestral gluconeogenic enzyme. Nature 464:1077–1081

Schafer DA, Schroer TA (1999) Actin-related proteins. Annu Rev Cell Dev Biol 15:341–363

Scheer U, Hinssen H, Franke WW, Jockusch BM (1984) Microinjection of actin-binding proteins and actin antibodies demonstrates involvement of nuclear actin in transcription of lampbrush chromosomes. Cell 39:111–122

Schleicher M, Jockusch BM (2008) Actin: its cumbersome pilgrimage through cellular compartments. Histochem Cell Biol 129:695–704

Schoenenberger CA, Buchmeier S, Boerries M, Sutterlin R, Aebi U, Jockusch BM (2005) Conformation-specific antibodies reveal distinct actin structures in the nucleus and the cytoplasm. J Struct Biol 152:157–168

Skarp KP, Vartiainen MK (2010) Actin on DNA-an ancient and dynamic relationship. Cytoskeleton 67:487–495

Van den Ent F, Amos LA, Lowe J (2001) Prokaryotic origin of the actin cytoskeleton. Nature 413:39–44

Van den Ent F, Moeller-Jensen J, Amos AL, Gerdes K, Löwe J (2002) F-actin-like filaments formed by plasmid segregation protein ParM. Embo J 21:6935–6943

Waidner B, Specht M, Dempwolff F, Haeberer K, Schaetzle S, Speth V, Kist M, Graumann PL (2009) A novel system of cytoskeletal elements in the human pathogen *Heliobacter pylori*. PLoS Pathog 5:e1000669

Wang S, Arellano-Santoyo H, Combs PA, Shaevitz JW (2010) Actin-like cytoskeleton filaments contribute to cell mechanics in bacteria. Proc Natl Acad Sci USA 107:9182–9185

Zhao YJ, Hoffmann-Rohrer U, Grummt I (2008) Nuclear myosin I acts in concert with polymeric actin to drive RNA polymerase I transcription. Genes Dev 22:322–330

Transient Receptor Potential Cation Channels in Pancreatic β Cells

Barbara Colsoul, Rudi Vennekens, and Bernd Nilius

Abstract There is now overwhelming evidence that TRP channels might play a significant role in the regulation of insulin release from pancreatic β cells, which is until now insufficiently recognized. TRP channels are abundantly expressed on β cells. The focus of this review will be on cation channels from the melastatin TRP subfamily. We will discuss how TRPM channels can influence Ca^{2+} signaling in β cells. Knock out models of TRPM2 and TRPM5, which show a pre-diabetic phenotype, will be illustrative for this purpose. Based on these insights, TRPM5 will be critically evaluated as a potential drug target for diabetes type II therapy, which has received currently a high interest of the pharmaceutical industry. In addition, an unexpected role of the TRP channel TRPM3 as a gatekeeper of zinc, which is required for insulin storage, will be considered. Finally, we will critically discuss the use of mouse models for the unraveling of basic mechanisms of insulin release. The study of the role of TRP channels in the regulation of insulin release is of wide interest for fundamental research, evaluation of molecular mechanisms of disease and exploration of novel drug targets for metabolic diseases.

1 Basic Mechanisms of Insulin Secretion in Beta Cells

The β cell is one of five types of cells present in the pancreatic islets of Langerhans, which are islands of cells distributed throughout the pancreas (Kulkarni 2004). The islet is a complex structure consisting of 5 different cell types: insulin-secreting β cells (which constitute 65–90% of the islet cell population), glucagon-releasing

B. Colsoul (✉)
KU Leuven, Department of Molecular Cell Biology,
Laboratory of Ion Channel Research, Campus Gasthuisberg,
Herestraat 49, bus 802, 3000, Leuven, Belgium
e-mail: barbara.colsoul@med.kuleuven.be

Rev Physiol Biochem Pharmacol, DOI: 10.1007/112_2011_2,

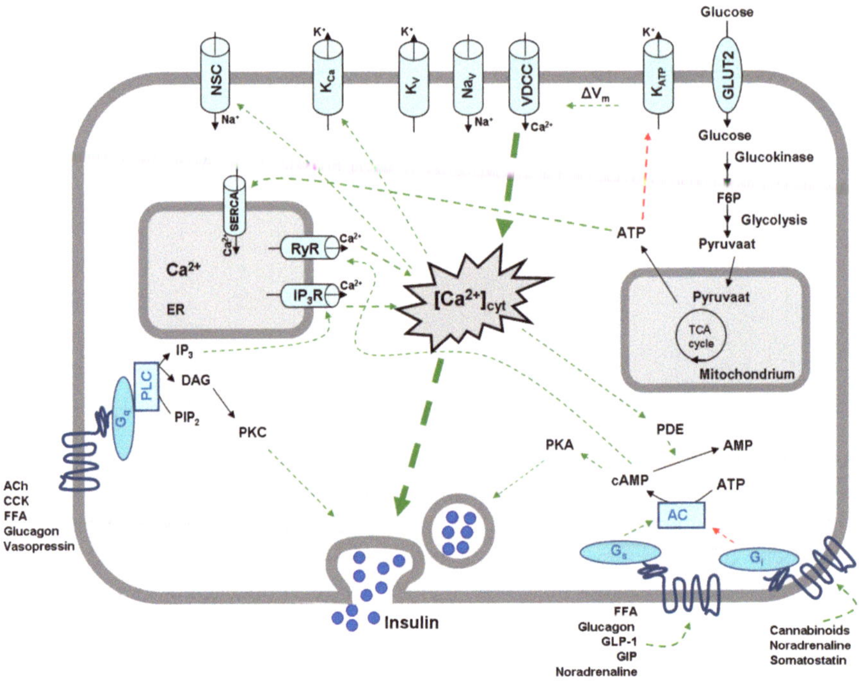

Fig. 1 Stimulus-secretion coupling of the pancreatic β cell. Glucose enters the beta cell and glucose metabolism produces ATP, which closes ATP-sensitive K^+ channels. This leads to a series of events with depolarization of the membrane potential and activation of voltage-dependent Ca^{2+} channels. The resulting Ca^{2+}-influx and increase in cytosolic Ca^{2+} triggers exocytosis of insulin-containing vesicles. Many other ion channels influence the electrical activity and the insulin release can be modulated by many factors, including hormones and neurotransmitters, as indicated. *NSC* non-selective cation channel, K_{Ca} Ca^{2+}-activated K^+ channels, K_V voltage-gated K^+ channels, Na_V voltage-gated Na^+ channels, *VDCC* voltage-dependent Ca^{2+} channel, K_{ATP} ATP-sensitive K^+ channel, *GLUT2* glucose transporter 2, *F6P* fructose-6-phosphate, *SERCA* sarco/endoplasmic reticulum Ca^{2+}-ATPase, IP_3R IP_3 receptor, *RyR* ryanodine receptor, *ER* endoplasmic reticulum, *PDE* phosphodiesterase, *AC* adenlyly cyclase, *FFA* free fatty acids, *GIP* glucose-dependent insulinotropic peptide, *cAMP* cyclic adenosine monophosphate

α cells, somatostatin-producing δ cells, polypeptide-containing PP-cells and ghrelin-secreting ε cells. Insulin is synthesized and secreted into the blood by the β cells mainly in response to glucose but also in response to other nutrients (such as amino acids and fatty acids), hormones (e.g. the incretin hormones GLP-1 and GIP) and neurotransmitters (e.g. ACh) (Newsholme et al. 2010; Winzell and Ahren 2007). Insulin lowers the blood glucose by promoting glucose uptake and nutrient storage in muscle, fat and liver. As it is the only blood glucose lowering hormone and essential for glucose homeostasis, its secretion is a very tightly regulated process that when altered may result in hypoglycemia or hyperglycemia. The latter can result in glucose intolerance and diabetes mellitus (LeRoith 2002).

The secretion of insulin by the pancreatic β cell is a complex process driven by electrical activity and oscillations of the intracellular Ca^{2+} concentration, $[Ca^{2+}]_i$ [see Fig. 1 and (MacDonald and Rorsman 2006)]. Briefly, glucose enters the β cell via the high K_m GLUT-2 transporter. The rate-limiting glucosensor that determines the characteristic sigmoidal glucose concentration dependence of insulin secretion is the first phosphorylation step in glucose metabolism exerted by the high K_m glucokinase. The metabolism of glucose produces ATP, which closes ATP-sensitive K^+ channels. Consequently, the membrane potential will depolarize, supported by a still unidentified depolarizing conductance (Henquin 2009; Henquin et al. 2009). Indeed, the closure of K_{ATP} increases the input resistance of the β cell, allowing e.g. small inward currents to generate significant depolarizations (Rorsman and Trube 1985). These mechanisms may result in the activation of voltage-dependent Ca^{2+} channels. The ensuing Ca^{2+} increase, together with the resulting Ca^{2+}-induced Ca^{2+} release (Lemmens et al. 2001), causes exocytosis of insulin-containing vesicles. Ca^{2+} stimulates insulin secretion by regulating docking and initiating fusion of secretory granules with the plasma membrane, a process mediated by SNARE proteins (MacDonald and Rorsman 2007).

2 Role of Ion Channels: The Classical View

The activity of ATP-sensitive K^+ channels, voltage-dependent Ca^{2+} channels and other ion channels results in a typical pattern of electrical activity of the beta cell, consisting of slow waves of depolarized plateaus on which bursts of action potentials are superimposed (Ashcroft and Rorsman 1989; Drews et al. 2010) and separated by electrically silent intervals. The oscillating process regenerates as long as the glucose concentration is elevated and is glucose-dependent: with increasing glucose concentration, burst phases are prolonged and interburst phases are shortened until continuous activity is reached at glucose concentrations above ~25 mM. This typical pattern of electrical activity is accompanied by simultaneous oscillatory changes in $[Ca^{2+}]_i$. The origin of this characteristic pattern is a complex interplay between different ion channels, intracellular Ca^{2+} levels ($[Ca^{2+}]_{cyt}$) and the cellular metabolism of the beta cell (Bertram et al. 2007). After closure of ATP-sensitive K^+ channels as a result of glucose metabolism and ATP production, the membrane potential depolarizes with the help of an unidentified depolarizing background conductance. The activation of voltage-dependent Ca^{2+} channels causes the initiation of the electrical activity and is responsible for the fast upstroke of the action potentials on top of the burst. The repolarizing phase of the action potential is caused by activation of delayed-rectifier and Ca^{2+}-dependent K^+ channels. These channels close at around −40 mV, which is the origin of the plateau potential (Ashcroft and Rorsman 1989).

Why do β cells generate an oscillatory pattern of $[Ca^{2+}]_i$ during constant glucose stimulation? The pattern results mainly from an equilibrium between hyperpolarizing K^+ conductances and depolarizing Ca^{2+} conductances and the following course of events has been proposed (Ashcroft and Rorsman 1989; Drews et al. 2010; MacDonald and Rorsman 2006). While the depolarizing component dominates at the beginning of the burst, this decreases during the burst of activity due to Ca^{2+}-dependent inactivation of VDCC. Furthermore, Ca^{2+}-dependent K^+ channels and ATP-sensitive K^+ channels will be activated during the burst as a result of the increase in intracellular Ca^{2+} and a lowering of the ATP/ADP ratio (due to an increased Ca^{2+}-ATPase activity and depolarization of the mitochondrial potential $\Delta\Psi$) respectively. As a result, the V_m hyperpolarizes and the burst terminates. During the interburst interval, the intracellular Ca^{2+} concentration lowers which results in a decrease of Ca^{2+}-dependent K^+ channel activity, a decrease in the activity of Ca^{2+}-ATPase, a hyperpolarization of the mitochondrial potential $\Delta\Psi$ and a resulting increase in ATP production and finally a recovery of the voltage-dependent Ca^{2+} channels from inactivation. All these events result in a depolarization of the membrane potential, an increase in intracellular Ca^{2+} concentration and the initiation of the next burst. The duration of the interburst interval is proposed to be determined by the rate of recovery of voltage-dependent Ca^{2+} channels from inactivation (Santos et al. 1991).

This glucose-induced electrical activity and insulin release can be modulated by several factors, including several hormones, neurotransmitters and nutrients like amino acids (see Fig. 1). The incretin hormone GLP-1 enhances glucose-induced insulin secretion and is suggested to depolarize the Vm in several ways, including closure of K_{ATP} channels, activation of yet unidentified nonselective cation currents and Ca^{2+}-mobilization from intracellular stores (Miura and Matsui 2003; Holst 2007; Winzell and Ahren 2007). The neurotransmitter ACh enhances insulin secretion by acting via the PLC pathway and by activation of a Na^+ current. The latter mechanism is still unclear, but might be due to activation of the NALCN leak channel via a G-protein independent pathway (Gilon and Rorsman 2009). Finally, the hormones glucagon and somatostatin, released by α and δ cells from the pancreatic islet, can respectively stimulate and inhibit insulin release in a paracrine manner.

3 Expression of TRP Channels in Pancreatic β Cells

The consensus model of glucose-induced insulin release attributes the depolarization of the membrane potential towards the initiation of electrical activity to the closure of ATP-sensitive K^+ channels. However, this event needs the presence of an additional inward background current that can bring the membrane potential away from the equilibrium potential of K^+. Because of the increased input resistance of the β cells by closure of K_{ATP}, only small currents are required to bring about depolarization. At this point, the identity of this current is still unclear (Gilon and Rorsman 2009; Henquin 2009; Henquin et al. 2009). In theory, the current could be produced by an efflux of Cl^- or a cation influx. TRP channels were often regarded as interesting candidates for this background depolarizing current (Drews et al. 2010; Jacobson and Philipson 2007).

Transient receptor potential (TRP) channels form a large group of cation channels with 28 mammalian members that can be subdivided into six main subfamilies based on amino acid homology (see Fig. 2): TRP canonical (TRPC; TRPC1-7, with *Trpc2* being a pseudo gene in humans), TRP vanilloid (TRPV; TRPV1-6), TRP melastatin (TRPM; TRPM1-8), TRP polycystin (TRPP; TRPP2, TRPP3 and

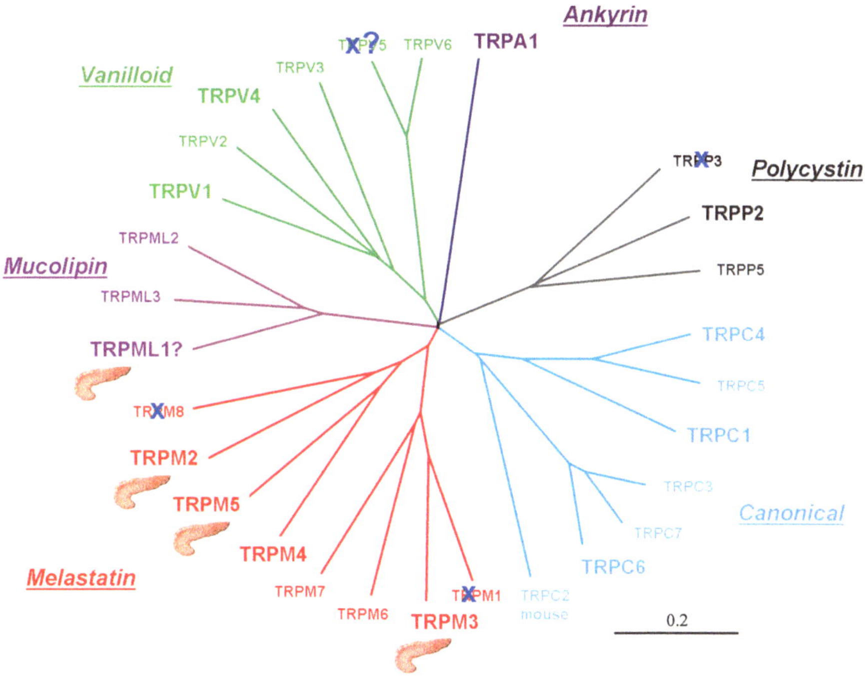

Fig. 2 Phylogenetic tree of the mammalian TRP superfamily. TRPC (canonical), TRPM (melastatin), TRPV (vanilloid), TRPA (ankyrin), TRPP (polycystin), and TRPML (mucolipin). *Trpc2* is a pseudogene in humans. TRP channels with a probably high functional impact are indicated by a *logo*, *larger letter type* indicates high expression but lack of a functional impact, *small letter type* indicates low expression. TRPM2, TRPM5 and TRPM3 are considered as the most important TRP channels in regulating insulin release. In our own assays, TRPML1 and TRPP2 are the channels with the highest expression (Colsoul B., Nilius B., Vennekens R. unpublished).The role of the highly expressed intracellular channel TRPML1 might be related to endo/exocytosis. TRP channels marked by a *cross* are probably not present in β cells

TRPP5), TRP mucolipin (TRPML; TRPML1-3) and TRP ankyrin (TRPA; TRPA1) (Pedersen et al. 2005). Topologically, all TRP channels are membrane proteins with six putative transmembrane segments (S1–S6) and a cation permeable pore-forming loop between S5 and S6. Most TRP channels are non-selective Ca^{2+} permeable channels, although the permeability ratios P_{Ca}/P_{Na} vary considerably, ranging from 0.3 for TRPM2 to >100 for TRPV5 and TRPV6 (Owsianik et al. 2006). TRPM4 and TRPM5 are the only two TRP channels that are impermeable to Ca^{2+} and exclusively permeable to monovalent cations. The gating of TRP channels is very diverse (Nilius et al. 2005a, 2007a; Rohacs and Nilius 2007; Voets and Nilius 2007) and TRP channels can be activated by a plethora of stimuli, including ligand binding, voltage, cell swelling or temperature. Due to this diversity of stimuli, TRP channels can act as polymodal cellular sensors in many cell types: they measure changes in the environment to initiate adequate cell, organ and behavioral responses, e.g. taste/pain transduction and temperature sensing. They contribute to changes in the cytosolic free Ca^{2+} concentration either by acting as Ca^{2+} entry pathways in the plasma membrane or by influencing the membrane potential, in this way modulating the driving force for Ca^{2+} entry mediated by alternative pathways. Thus, TRP channels influence a plethora of Ca^{2+}-dependent cell functions such as gene transcription, migration, cell death and exocytosis and are implicated in human diseases (Nilius et al. 2005b, 2007b).

Interestingly, more and more TRP channels are being described in primary beta cells or insulin-secreting cell lines [see Table 1, Fig. 1 and (Jacobson and Philipson 2007)]. Indeed, TRPV2, TRPV4, TRPC1, TRPC4, TRPC6, TRPM2, TRPM3, TRPM4 and TRPM5 are identified in insulinoma cell lines such as MIN6 (mouse) or INS-1 (rat). Many of these channels have also been described in primary tissue with expression of TRPV2, TRPV4, TRPC1, TRPM2, TRPM3, TRPM4 and TRPM5 reported in mouse islets and TRPM2, TRPV5, TRPC1 and TRPC4 in rat islets or beta cells. Interestingly, no expression could be found in mouse islets for TRPV5 [which contrasts with earlier reports demonstrating expression in rat islets (Janssen et al. 2002)], TRPM1, TRPM8 or TRPP3, whereas TRPML1 and TRPP2 are highly expressed (Colsoul B., Nilius B., Vennekens R., unpublished). Not much is known yet about expression of TRP channels in human tissue: TRPV5 and TRPV6 are detected in human pancreas and TRPM2, TRPM4 and TRPM5 are specifically reported to be expressed in human islets. In the next sections, we describe the current knowledge of the function of these TRP channels in insulin release.

4 Modulation of Insulin Release by TRP Channels

4.1 TRP Canonicals

TRPC channels are non-selective Ca^{2+} permeable cation channels, with the selectivity ratio P_{Ca}/P_{Na} varying significantly between the different family members (Gees et al. 2011). TRPC channels are widely expressed and their characterization is complicated by the possible occurrence of heterotetramers. TRPC1 mRNA could be detected in INS-1 cells and rat beta cells and at high levels in mouse islets and MIN6 cells, whereas it could not be detected in another mouse insulinoma cell line βTC3. (Sakura and Ashcroft 1997; Li and Zhang 2009). Four transcripts for mouse TRPC1 RNA, representing different splice variants, have been found in mouse islets and the beta cell line MIN6 (Sakura and Ashcroft 1997). However, the influence of TRPC1 on insulin release still needs to be determined.

Not much is known about other TRPC channels in insulin release. Although TRPC4 could be detected in βTC3 and INS-1 cells and in rat beta cells and mouse islets (Roe et al. 1998; Freichel et al. 2004; Li and Zhang 2009), analysis of blood glucose homeostasis by glucose tolerance tests did not reveal differences between WT and *Trpc4*-deficient mice both regarding basal glucose levels under fasting conditions as well as following intraperitoneal glucose challenge (Freichel et al. 2004). Since TRPC4 is activated by phospolipase C pathways, it cannot be excluded that

Table 1 Overview of the TRP channels that are shown to be expressed in endocrine pancreas and their proposed function

TRP	Expression in pancreas	Method	General features and activation mechanisms	Proposed function in β cells	Reference
TRPV1	Controversy: expression shown in Sprague-Dawley rat islets and β cell lines (RIN and INS-1); no expression detected in mouse islets or ZDF rat islets	RT-PCR; western blot; immuno-fluorescence	Heat (≥43°C), capsaicin	Unknown	Caterina et al. (1997), Hayes et al. (2000), Akiba et al. (2004), Voets et al. (2004), Razavi et al. (2006), Gram et al. (2007)
TRPV2	MIN6; mouse islets	RT-PCR	Noxious heat (>53°C); translocation to plasma membrane by IGF-1 and insulin	Insulin-induced enhancement of insulin secretion	Caterina et al. (1999), Beech et al. (2004), Hisanaga et al. (2009), Vennekens et al. (2008)
TRPV4	MIN6; mouse pancreas	RT-PCR	Moderate heat (>23°C); cell swelling; 4α-PDD; EETs; shear stress, arachidonic acid	hIAPP-induced Ca^{2+} elevation and apoptosis	Chung et al. (2003), Suzuki et al. (2003), Casas et al. (2008), Masuyama et al. (2008), Everaerts et al. (2010)
TRPV5	Rat pancreatic islets (localization to secretory granules); human pancreas	RT-PCR; immunohis-tochemistry	Highly Ca^{2+}-selective; constitutively active	Ca^{2+} efflux pathway from secretory granules	Muller et al. (2000), Hoenderop et al. (2001, 2003, 2005), Janssen et al. (2002), Nijenhuis et al. (2003a)
TRPV6	Human pancreas	RT-PCR	Highly Ca^{2+}-selective; constitutively active	Unknown	Muller et al. (2000), Hoenderop et al. (2001, 2003, 2005), Nijenhuis et al. (2003a, b)
TRPC1	MIN6; mouse islets; INS-1; rat beta cells (no detection in βTC-3)	Northern blot; RT-PCR	G_q-PLC pathways (PLC); store depletion	Unknown	Sakura and Ashcroft (1997), Roe et al. (1998), Hofmann et al. (2000), Kim et al. (2003), Liu et al. (2003), Rychkov and Barritt (2007), Li and Zhang (2009)

(continued)

Table 1 (continued)

TRP	Expression in pancreas	Method	General features and activation mechanisms	Proposed function in β cells	Reference
TRPC4	βTC-3; INS-1; rat beta cells; mouse islets	Northern blot; RT-PCR	G_q-PLC pathways	Unknown	Roe et al. (1998), Freichel et al. (2001, 2004), Qian et al. (2002), Tiruppathi et al. (2002), Cavalie (2007), Li and Zhang (2009)
TRPC6	βTC-3	Northern blot	G_q-PLC pathways (DAG)	Unknown	Roe et al. (1998), Hassock et al. (2002), Estacion et al. (2004), Li and Zhang (2009)
TRPM2	Several cell lines; rat beta cells; human and mouse islets (localization to plasma membrane and lysosomes)	RT-PCR; immunoblot; immunofluorescence	ADP-ribose; cADPR; NAD; heat; H_2O_2; Ca^{2+}	Glucose-, heat- and GLP-1-induced insulin release; H_2O_2-mediated cell death	Uchida et al. (2011), Qian et al. (2002), Inamura et al. (2003), Kraft and Harteneck (2005), Kuhn et al. (2005), Togashi et al. (2006), Lange et al. (2009), Li and Zhang (2009)
TRPM3	INS-1; mouse islets	RT-PCR; northern blot; western blot	Steroid hormones (PS); hypotonic cell swelling	PS-induced Ca^{2+}-increase; zinc influx	Wagner et al. (2010), Grimm et al. (2003), Lee et al. (2003), Wagner et al. (2008)
TRPM4	Several beta cell lines; mouse and human islets	RT-PCR, immunoprecipitation; western blot; immunofluorescence	Ca^{2+}-impermeable; activation by IC Ca^{2+}, PKC, heat, PIP_2; inhibition by ATP; voltage dependent	Controversy	Cheng et al. (2007), Vennekens et al. (2007), Marigo et al. (2009), Launay et al. (2002), Ullrich et al. (2005)
TRPM5	INS-1; MIN6; mouse and human islets	RT-PCR; immunofluorescence	Ca^{2+}-impermeable; activation by IC Ca^{2+}, heat, PIP_2; voltage dependent	Positive regulation of glucose-stimulated insulin secretion	Brixel et al. (2010), Colsoul et al. (2010), Hofmann et al. (2003), Liu and Liman (2003), Prawitt et al. (2003), Ullrich et al. (2005)

the channel is involved in ACh- or glucagon-induced amplification of insulin release. Finally, TRPC6 transcripts have been detected, although to a low level, in βTC-3 insulin-secreting cells. Thus, it is clear that more research is needed in order to clarify the possible function of TRPC channels in the insulin release of the pancreatic beta cell. It has been suggested that TRPC channels mediate the unknown depolarizing current that account for the Ca^{2+}-release activated cation current characterized earlier in βTC3 cells (Roe et al. 1998).

4.2 *TRPM2*

TRPM2 is a non-selective Ca^{2+} permeable cation channel with unique gating properties that is fused C-terminally to an enzymatic ADP-ribose pyrophosphatase domain (Perraud et al. 2001). The channel is shown to be expressed in insulin-secreting cell lines, such as the rat cell lines CRI-G1 and RIN-5F, and in human and mouse pancreatic islets (Hara et al. 2002; Qian et al. 2002; Inamura et al. 2003; Togashi et al. 2006). Moreover, the channel co-expresses with insulin but not with glucagon, indicating expression in beta cells (Togashi et al. 2006).

TRPM2 is shown to be activated by various stimuli, including adenine dinucleotides (ADPR, cADPR, NAADP, β-NAD), reactive oxygen species (ROS) such as H_2O_2 and OH^-, and intracellular Ca^{2+} (Perraud et al. 2001; Sano et al. 2001; Hara et al. 2002; Du et al. 2009). In the rat insulinoma cell line CRI-G1, a current with TRPM2-like properties (such as the need of intracellular Ca^{2+} for current activation, an intermediate Ca^{2+} permeability ratio and activation by ADPR and β-NAD) could be detected (Inamura et al. 2003). Furthermore, ADPR elicited rapid activation of linear currents with biophysical and pharmacological characteristics typical of TRPM2 in INS-1 cells and primary mouse beta cells. In *Trpm2*$^{-/-}$ primary beta cells, no ADPR-elicited current could be detected, suggesting that TRPM2 is natively expressed and forms a functional channel in beta cells (Lange et al. 2009).

TRPM2 is important in beta cell apoptosis, a feature linked to the activation of the channel by H_2O_2 and OH^-. These ROS, that are produced by oxidative stress, are thought to play a central role in beta cell death and the consequent development of type 1 and type 2 diabetes (Mandrup-Poulsen 2003; Rhodes 2005). Indeed, activation of TRPM2 by H_2O_2 has been shown to mediate Ca^{2+} influx and beta cell death in a rat beta cell line RIN-5F that natively expresses TRPM2 (Hara et al. 2002; Ishii et al. 2006). Moreover, INS-1 cells with suppressed TRPM2 expression are 72% less affected by H_2O_2-induced cell death (Lange et al. 2009). The H_2O_2-induced Ca^{2+} influx is thought to be mediated by increasing levels of NAD^+, that binds directly to the Nudix motif in the cytosolic C terminal of TRPM2 (Hara et al. 2002). Furthermore, the H_2O_2-induced Ca^{2+} increase is reduced when the cells are treated with scavengers of the hydroxyl radical (Ishii et al. 2006), indicating that H_2O_2 acts by generation of free radicals in the cell interior. TRPM2 activation by H_2O_2 can also be mediated by release of ADP-ribose, a metabolite of NAD^+, from the mitochondria. ADP-ribose directly binds the Nudix motif and in this way activates TRPM2 (Perraud et al. 2001).

Next to the involvement in oxidative stress sensing, TRPM2 has been suggested to contribute to insulin release induced by heat, glucose and incretin hormones (Uchida et al. 2011; Togashi et al. 2006). Indeed, forskolin- (an activator of adenylyl cyclase) and exendin-4- (a GLP-1 receptor agonist) induced insulin release from rat pancreatic islets was significantly reduced in si-TRPM2-treated islets (Togashi et al. 2006). Furthermore, 2-aminoethoxydiphenyl borate (2-APB), a rapid and reversible inhibitor of TRPM2, inhibits both heat- and exendin-4- evoked insulin release from rat pancreatic islets (Togashi et al. 2008). These pharmacological data are confirmed in studies using the *Trpm2*$^{-/-}$ mouse: insulin secretion induced by glucose and GLP-1 was seriously impaired in *Trpm2*-deficient islets, whereas the response to tolbutamide, a K_{ATP} channel inhibitor, was unchanged (Uchida et al. 2011). This impairment of insulin secretion is caused by reduced increases in intracellular Ca^{2+}, indicating that TRPM2 mediates Ca^{2+} influx during glucose- and/or GLP-1 stimulation. However, the situation might be more complex, since glucose-stimulated insulin secretion evoked under conditions of glucose, diazoxide and high K^+ (conditions designed to "clamp" intracellular Ca^{2+}), was lost in *Trpm2*-deficient islets. Since the intracellular Ca^{2+} under these conditions was not altered, these data suggest that TRPM2 mediates insulin secretion independent of its role as a Ca^{2+} entry channel (Uchida et al. 2011).

Finally, TRPM2 has been reported to have an additional role as an intracellular Ca^{2+} release channel in pancreatic beta cells (Lange et al. 2009). Indeed, internally applied ADPR gives rise to a single Ca^{2+} transient both in INS-1 and in primary mouse beta cells and this effect was completely abolished in *Trpm2*$^{-/-}$ primary mouse beta cells. Furthermore, TRPM2 colocalizes with lysosome-associated membrane protein-1 (LAMP1), a specific marker for lysosomes. In agreement with this, ADPR-induced intracellular Ca^{2+} release was abolished in INS-1 cells treated with bafilomycin A, a macrolide antibiotic that empties lysosomal Ca^{2+} stores without affecting ER stores (Bowman et al. 1988). These data indicate that ADPR-dependent TRPM2-mediated Ca^{2+} release occurs predominantly from a lysosomal store. In addition, it is suggested that TRPM2-mediated Ca^{2+} release contributes to H_2O_2-induced apoptosis (Lange et al. 2009). Indeed, H_2O_2 induces significant cell death in INS-1 cells in the absence of extracellular Ca^{2+}, albeit with a reduced severity. This effect was reduced in cells with reduced TRPM2 expression, indicating that not only Ca^{2+} influx through plasma membrane TRPM2 but also TRPM2-dependent lysosomal Ca^{2+} release plays a critical role in H_2O_2-mediated beta cell death (Lange et al. 2009).

4.3 TRPM3

TRPM3 is a member of the melastatin subfamily of TRP channels with limited homology to the heat-sensitive TRPV channels. The TRPM3 gene encodes for the largest number of different TRPM3 isoforms due to alternative splicing and exon usage, leading to channels with divergent pore and gating properties (Oberwinkler et al. 2005). Interestingly, TRPM3 is so far the only TRP channel which hosts a microRNA gene (intron 8 encodes miR-204), which may regulate a variety of target genes at the transcriptional level (Weber 2005; Oberwinkler and Phillipp 2007). Theoretically, the number of possible variants due to alternative splicing is indeed huge. We refer here only to the (probably most common) variant TRPM3α2 (1709 amino acids: the pore lacks 12 aa in comparison to the longest form TRPM3α1 of 1721 aa).

TRPM3 is expressed in a variety of neuronal and non-neuronal tissue (Oberwinkler and Phillipp 2007), including whole pancreas (Grimm et al. 2003; Fonfria et al. 2006), INS-1 cells and mouse pancreatic islets (Klose et al. 2011; Wagner et al. 2008). TRPM3 channels are directly activated by the neuro-steroid hormone pregnenolone-sulphate (PS). Pancreatic beta cells and INS-1 cells express PS-sensitive channels that share several pharmacological and biophysical properties of recombinant TRPM3 channels [such as sensitivity to nifedipine and block by monovalent cations (Wagner et al. 2008)]. Moreover, PS elicits a large Ca^{2+} increase in INS-1 cells and pancreatic islets, an action dependent on TRPM3 expression. This PS-induced Ca^{2+} increase could be blocked by the selective and potent TRPM3 blocker mefenamic acid in INS-1E cells and mouse pancreatic islets (Klose et al. 2011). Remarkably, mefenamic acid did not block glucose- or tolbutamide- induced Ca^{2+} increase, indicating that TRPM3 is not involved in the K_{ATP}-dependent Ca^{2+} signaling of the beta cell. PS did increase however glucose-induced insulin secretion from pancreatic islets, an effect abolished by mefenamic acid (Klose et al. 2011; Wagner et al. 2008). Thus, TRPM3 is present in pancreatic β cells and activation of the channel increases insulin release via intracellular Ca^{2+} increase although the channel does not seem to influence glucose-induced insulin secretion. Interestingly, PS activation of β cells (via TRPM3 and voltage-gated Ca^{2+} channels) induces the biosynthesis of a gene regulatory protein, the zinc finger transcription factor Egr-1, and in this way leads to increased biosynthesis of insulin (Mayer et al. 2011). However, as the pharmacological concentrations of PS (50 μM) used to demonstrate enhancement of insulin secretion do not occur in vivo, it seems unlikely that PS has an important role in the direct regulation of insulin secretion. It is possible that TRPM3 plays a role in certain conditions where elevated plasma PS levels and changes in glucose homeostasis co-occur (like pregnancy or 21-hydroxylase-deficiency).

4.4 TRPM4

TRPM4 is a Ca^{2+}-activated nonselective monovalent cation channel that is impermeable to divalent cations. The channel has been proposed to control insulin secretion in a rat insulinoma cell line INS-1, where TRPM4 protein is abundantly expressed (Cheng et al. 2007). TRPM4 expression and TRPM4-like channel activity could be detected in the beta cell lines INS-1, HIT-T15, RINm5F, β-TC3 and MIN-6 and the alpha cell line INR1G9 (Cheng et al. 2007; Marigo et al. 2009). Inhibition of TRPM4 decreases the magnitude of the Ca^{2+} signal and insulin release in response to glucose, AVP (arginine-vasopressin, a Gq-coupled receptor agonist in β-cells) and glyburide (glibenclamide) in INS-1 cells (Cheng et al. 2007; Marigo et al. 2009). These data suggest that depolarizing currents generated by TRPM4 are an important component in the control of intracellular Ca^{2+} signals necessary for insulin secretion. Furthermore, it is suggested that TRPM4-containing vesicles are translocated to the plasma membrane via Ca^{2+}-dependent exocytosis, which may represent a regulatory mechanism by which beta cells regulate electrical activity (Cheng et al. 2007; Marigo et al. 2009). However, all these studies have been performed on cell lines. Although TRPM4 protein expression could be found within insulin-producing human beta cells and mouse pancreatic islets (Marigo et al. 2009), studies on *Trpm4*$^{-/-}$ mice revealed no difference in glucose-induced insulin secretion from freshly isolated pancreatic islets (Vennekens et al. 2007). Moreover, these mice did not suffer from an impaired glucose tolerance after an intraperitoneal injection of glucose. These data suggest that TRPM4 is probably not involved in the signal mechanism following glucose stimulation. On the other hand, this does not exclude a possible role for TRPM4 in G_q- or G_s-receptor-coupled signaling pathways, for example during stimulation with e.g. glucagon or GLP-1. Additionally, TRPM4 is proposed to be involved in glucagon secretion from the pancreatic alpha cell line αTC1-6 (Nelson et al. 2011): TRPM4 inhibition decreased the magnitude of intracellular Ca^{2+} signals and glucagon secretion in response to several agonists such as the G_q–protein coupled receptor agonist AVP and high K^+ (Nelson et al. 2011).

5 TRPM5 as a Modulator of Ca^{2+} Oscillations

TRPM5, like its closest homologue TRPM4, is a Ca^{2+}-activated nonselective monovalent cation channel that is impermeable to divalent cations (Hofmann et al. 2003). The channel is expressed in β cells from pancreatic islets, as its expression co-localizes with insulin. A Ca^{2+}-activated non-selective monovalent cation channel could be measured in β cells and was largely reduced in *Trpm5*$^{-/-}$ mice, indicating that TRPM5 is an important constituent of the Ca^{2+}-activated cation current in β cells (Colsoul et al. 2010). The TRPM5-dependent current is of a small magnitude, ~2pA/pF at −80 mV and 1.5 μM $[Ca^{2+}]_i$ (Colsoul et al. 2010), suggesting that the channel can only influence the electrical activity of β cells under conditions of high electrical resistance. Indeed, whereas no difference could be detected in V_m or intracellular Ca^{2+} in non-stimulatory (low glucose, high K_{ATP} activity) conditions of low electrical resistance, TRPM5 seems to influence electrical activity during glucose stimulation (a condition with low K_{ATP} activity and consequently a high electrical resistance). Whereas normal wild type islets respond to glucose stimulation with three types of oscillations (slow, mixed or fast), *Trpm5*$^{-/-}$ islets displayed specifically a lack of fast glucose-induced oscillations in V_m and Ca^{2+} (Colsoul et al. 2010). TRPM5 contributes to the slow depolarization in the slow interburst interval of the glucose-induced electrical activity, in this way shortening the interburst interval and leading to faster glucose-induced oscillations in V_m and Ca^{2+}. Why TRPM5 is only functionally relevant in a (fast-oscillating) subpopulation of the islets remains unclear but it might be that the weight of TRPM5-mediated depolarization is coupled to the glycolytic rate in the cell. According to the dual oscillator model (Bertram et al. 2007), fast oscillations are characterized by a high glycolytic rate (and a resulting high ATP production). This situation would make TRPM5 activity able to depolarize V_m in the interburst interval, as the hyperpolarizing K_{ATP} current is largely inactive at that point. This is in contrast with the situation in slow oscillating islets, where the oscillating glycolytic rate and the resulting high activity of K_{ATP} in the interburst interval would make TRPM5 insufficient to depolarize V_m. Fast Ca^{2+} oscillations are shown to be more efficient than slow oscillations in triggering exocytosis of secretory vesicles and insulin release (Berggren et al. 2004). In line with this, glucose-induced insulin release was reduced in isolated pancreatic islets from *Trpm5*$^{-/-}$ mice.

6 Lessons Form Knock-Out Models: TRPM2 and TRPM5

It is clear that many TRP channels are proposed to be involved in insulin release, but the majority of evidence comes from insulinoma cell lines. In order to define the role of TRP channels in insulin release and glucose homeostasis, it is necessary to study their function in vivo. So far, 2 knock out animals display dysfunctions in their glucose homeostasis. Studies on the *Trpm2*$^{-/-}$ mouse confirmed that TRPM2 is involved in insulin secretion stimulated by glucose and that further potentiated by incretins (Uchida et al. 2011). The *Trpm2*$^{-/-}$ mice exhibit higher basal glucose levels and an impaired glucose tolerance caused by lower plasma insulin levels (Uchida et al. 2011). These data suggest that TRPM2 might be a new target for diabetes therapy. Two independent laboratories have described a similar phenotype for the *Trpm5*$^{-/-}$ mice. TRPM5 is involved in glucose-induced insulin release and *Trpm5*-deficient mice display an impaired glucose tolerance during oral and intraperitoneal glucose tolerance tests (Brixel et al. 2010; Colsoul et al. 2010), caused by lower plasma insulin levels. These data suggest that *Trpm2*- and *Trpm5*- deficient mice display a prediabetic phenotype, caused by beta cell dysfunction. The relevance of this prediabetic phenotype during conditions of higher insulin demand (such as pregnancy, obesitas, ageing, etc.) remains to be shown.

7 An Intestinal TRPM5 Connection: Incretins

Interestingly, TRPM5 might influence glucose homeostasis indirectly via the incretin effect. The channel is detected in enteroendocrine cells of the gut, a population of intestinal solitary cells that is believed to play a role in chemosensation and releases hormones such as glucagon-like peptide 1 (GLP-1). GLP-1 is released from L cells and is, together with glucose-dependent insulinotropic hormone (GIP) released from K cells, responsible for the so-called incretin effect: an increase in the release of insulin even before plasma glucose levels have increased (Burcelin 2005). Interestingly, enteroendocrine cells show similarities to the taste cells of the taste bud. The function of TRPM5 in the taste transduction of sweet, bitter and umami tastes is well described (Zhang et al. 2003). Binding of taste compounds to

GPCR in these cells results in activation of the GTP-binding protein α-gustducin, functional interaction with PLCβ2, formation of IP_3 and release of Ca^{2+} from intracellular stores. Consequently, TRPM5 is activated by the resulting increase in intracellular Ca^{2+} and causes depolarization of the membrane potential (Huang and Roper 2010). It is hypothesized that a similar pathway leads to GLP-1 release in enteroendocrine cells. Several reports describe expression of TRPM5 and other taste signaling molecules (taste receptors, α-gustducin and PCLβ2) in enteroendocrine cells and involvement of this pathway in the release of GLP-1 (Jang et al. 2007; Rozengurt and Sternini 2007; Kokrashvili et al. 2009b). E.g. *α-gustducin*$^{-/-}$ mice showed no significant rise in plasma concentrations of GLP-1 and a delayed rise in plasma insulin after glucose gavage (Jang et al. 2007; Kokrashvili et al. 2009a, b). Next, *T1r3*-(sweet taste receptor) deficient mice failed to release GLP-1 from their duodenal cells after glucose injection into the intestine. These data indicate that sweet receptors in enteroendocrine cells couple to heterotrimeric gustducin to detect extracellular glucose or other components and respond with secretion of GLP-1. The exact role of TRPM5 in this process has to our knowledge not been investigated yet. However, since the channel is expressed in enteroendocrine cells, it is conceivable that it plays a role in the sweet tastant-induced release of GLP-1 or other hormones. Bitter tastants are also shown to release gastrointestinal hormones such as CCK, GLP-1 and ghrelin (Janssen et al. 2011; Chen et al. 2006; Jeon et al. 2008), but the functional impact on insulin release or glucose homeostasis (and a possible role for TRPM5) remains to be shown (Fig. 3).

8 TRPM3: A Zinc Connection

As mentioned above, all TRP channels are cation channels with varying permeabilities for Ca^{2+} and Na^{+}. Most of these channels are important in different cell types either as Ca^{2+} channels or as Na^{+} entry channels and activation of TRP channels causes membrane depolarization and/or Ca^{2+} entry. However, TRPM3 channels are proposed to constitute a regulated Zn^{2+} entry pathway in pancreatic beta cells (Wagner et al. 2010). Zinc is important for insulin release as it is packed into co-crystals with insulin in the exocytotic vesicles. The formation of insulin crystals in beta cells depends amongst others on the ZnT8 transporter, which contributes to the packaging efficiency of stored insulin (Lemaire et al. 2009). Since Zn^{2+} ions are co-released with insulin, pancreatic beta cells need to continuously replenish their Zn^{2+} stores by taking up Zn^{2+} ions from the extracellular space. Insufficient Zn^{2+} uptake may lead to impaired insulin synthesis and aggravate diabetic symptoms (Chausmer 1998). Interestingly, TRPM3 channels in beta cells have been shown to be highly permeable for Zn^{2+} and capable of mediating Zn^{2+} uptake under physiological conditions through their own ion-conducting pore. The depolarization caused by the activation of TRPM3 channels leads to the activation of voltage-dependent Ca^{2+} channels and increases the Zn^{2+} influx through these channels (Gyulkhandanyan et al. 2006; Wagner et al. 2010). Thus, TRPM3 channels provide a novel regulated Zn^{2+} entry pathway in β cells and might in this way contribute to an improved insulin synthesis.

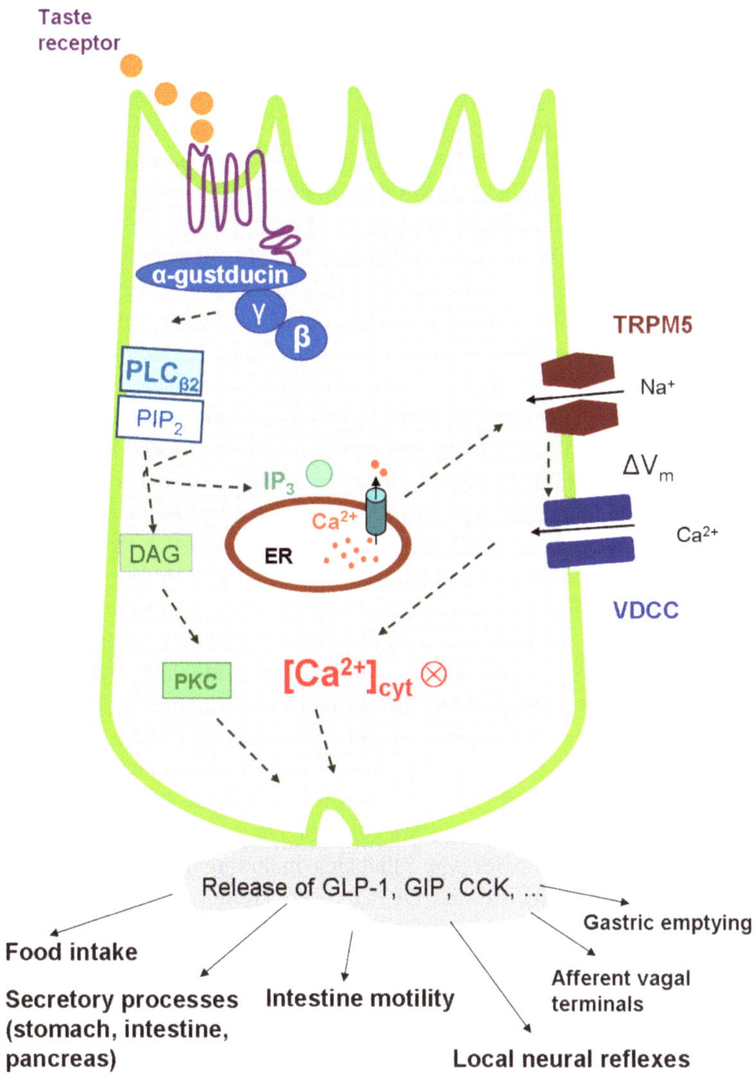

Fig. 3 Putative pathways triggered by sweet taste stimuli in gastrointestinal enteroendocrine cells. Sugar and other nutrients bind to sweet taste receptors in the apical portion of the enteroendocrine cell. This engages α-gustducin and PLCβ2, leading to intracellular Ca^{2+} release, activation of TRPM5, membrane depolarization and activation of voltage-dependent Ca^{2+} channels (VDCC). The resulting increase in intracellular Ca^{2+}, in combination with DAG-mediated activation of PKC, triggers the release of gastro-intestinal peptides such as GLP-1 and CCK. These peptides act in a paracrine or endocrine manner in order to exert diverse effects, among which amplification of glucose-induced insulin release. See also (Rozengurt and Sternini 2007)

9 TRP Channels and a Possible Role in Diabetes Type II

Disorders of glucose homeostasis can lead to diabetes mellitus, a group of metabolic diseases with the common feature hyperglycemia. Type 1 diabetes, the so-called "Insulin-Dependent Diabetes Mellitus", is caused by an autoimmune destruction of the β cells which subsequently leads to a lack of insulin production. This is in striking contrast to type 2 diabetes, where the β cells can still produce and release insulin, but cannot compensate for the increased demand due to e.g. ageing or obesity. Type 2 diabetes is a polygenic disease, acknowledged to result from a combination between peripheral insulin resistance and a genetically determined susceptibility to β cell dysfunction. It has been proposed that individuals at risk of type 2 diabetes carry one or more polymorphisms in genes encoding ion channels or in genes that regulate ion channel function, membrane targeting or expression, resulting in small changes in β cells electrical activity and consequently in reduced insulin secretion (Ashcroft and Rorsman 2004). Indeed, mutations in K_{ATP} channel genes are not only shown to result in congenital hyperinsulinism and neonatal diabetes, but also suggested to be associated with type 2 diabetes mellitus (Ashcroft 2005; Hattersley and Ashcroft 2005; Proks et al. 2005; Chistiakov et al. 2009). In this regard, it might be that gene variations of TRP channels known to influence insulin release in animal models play a role in the pathophysiology of type 2 diabetes mellitus. One study reports a negative association of beta cell function with three TRPM2 variants (rs2838553, rs2838554 and rs4818917) although no evidence for an association with type 2 diabetes mellitus could be found (Romero et al. 2010).

Interestingly, *Trpm5* expression is negatively correlated with blood glucose concentrations in the small intestine from diabetic patients (Young et al. 2009). Moreover, a recent study reports an association of TRPM5 variants with prediabetic phenotypes in subjects at risk for type 2 diabetes, including insulin secretion, insulin sensitivity and plasma glucose and GLP-1 levels (Ketterer et al. 2011). Indeed, TRPM5 SNP rs2301699 was significantly associated with insulin secretion and associated with lower GLP-1 levels during an oral glucose tolerance test. Furthermore, three TRPM5 SNPs (rs800344, rs800345 and rs2301699) were significantly associated with glucose levels during OGTT. Surprisingly, these SNPs were also associated with OGTT-derived insulin sensitivity (Ketterer et al. 2011). How these TRPM5 variants might affect insulin sensitivity remains elusive, since *Trpm5*$^{-/-}$ mice showed a normal insulin tolerance test (Colsoul et al. 2010). Furthermore, the functional impact of these mutations on the TRPM5 channel has not been clarified yet. However, these data indicate a possible link between TRPM5 and type 2 diabetes mellitus.

10 TRPM5 as a Drug Target

The involvement of TRPM5 in the insulin release and its link with type 2 diabetes suggest that TRPM5 might be a good drug target in the treatment of type 2 diabetes mellitus. A disadvantage of sulphonylureas (a frequently used medicine against diabetes mellitus type 2) is that, by closing the K_{ATP}-channels regardless of the glucose concentration, they bypass the normal metabolic regulation of the beta cell. Thus, insulin secretion continues even when plasma glucose has fallen to abnormally and dangerously low levels. TRPM5 activators would only increase insulin release at high blood glucose levels and in this way avoid the risk of hypoglycemia attacks. As TRPM5 is also implicated in transduction of sweet taste, activators of TRPM5 might be expected to increase sweet sensation and in this way reduce intake of e.g. sugars. Moreover, GLP-1 secretion might increase as a result of activation of TRPM5, resulting in an amplification of insulin release. Thus, an activator of TRPM5 has many potential benefits. However, this requires selective and efficient modulators of TRPM5. So far, only blockers of TRPM5 have been described. Flufenamic acid (FFA), a non-steroidal anti-inflammatory agent, is shown to inhibit TRPM5 but the compound is non-specific since several other cation channels (e.g. the closest homologue TRPM4) are also inhibited by FFA (Ullrich et al. 2005). Recently, triphenylphosphine oxide (TPPO) was identified as a selective and potent inhibitor of TRPM5 (Palmer et al. 2010). Indeed, it inhibits both human and mouse TRPM5 with an IC_{50} of 12 and 30 μM respectively and has no effect on TRPA1, TRPV1 or TRPM4. Clearly, the functional impact of all these TRPM5 modulators (and possible newcomers) should be critically tested in insulin release assays in isolated islets as well as in vivo (OGTT, IPGTT).

11 A Critical View: From Mouse to Man

In spite of the growing evidence of the function of TRP channels in mouse islets, no functional role for any TRP channel in human islets has been shown yet. What's more, expression data of TRP channels in human pancreatic islets and beta cells are often lacking. So far, TRPV5 and TRPV6 are detected in human pancreas (without any specification about expression in endocrine or exocrine cells) (Muller et al. 2000) and expression is shown specifically in human islets for TRPM2, TRPM4 and TRPM5 (Prawitt et al. 2003; Marigo et al. 2009). Concerning a functional role of these channels in nutrient-induced Ca^{2+} signalling and stimulus-secretion coupling, one can only speculate.

It is evident that human and mouse β-cells differ in many respects and that data obtained from rodent islets cannot be extrapolated to human islets without verification. E.g. whereas voltage-gated Na^{+} channels are inactive in rodent islets at physiological potentials and play no role in the bursting behavior, $Na_v1.6$ and $Na_v1.7$ channels are shown to participate in the action potential shaping during intermediate glucose-induced signaling and insulin secretion in human β cells (Barnett et al. 1995; Asensio et al. 2004). Human islets have been shown to respond to glucose and sulphonlyureas with Ca^{2+} oscillations that are synchronous throughout the islet and whose duration is modulated by glucose (Kindmark et al. 1994; Martin and Soria 1996). Even subjects with impaired glucose tolerance had islets that responded with oscillations in intracellular Ca^{2+} upon glucose stimulation (Kindmark et al. 1994). A detailed characterization of voltage-gated ion channels in human islets from non-diabetic donors revealed that voltage-gated T-type and L-type Ca^{2+} channels as well as Na^+ channels participate in glucose-stimulated electrical activity and insulin secretion, whereas Ca^{2+}-activated BK channels are required for rapid membrane repolarization and Ca^{2+}-influx through P/Q-type Ca^{2+} channels triggers exocytosis of insulin-containing granules (Braun et al. 2008). This is in many aspects different from the situation in rodent islets, confirming the fact that precaution must be taken when extrapolating data.

The study of ion channels in human islets requires selective and potent blockers. This is problematic for TRP channels, as many of the compounds known to modulate these channels also influence other ion channels. Furthermore, TRP channels would be expected to modulate insulin secretion in a subtle manner, a feature that makes the investigation towards their function in human islets a challenging task. However, as these channels are implicated in many human diseases and channelopathies (Nilius and Owsianik 2010; Nilius et al. 2007b), it is not inconceivable that mutations or dysfunctions of TRP channels contribute to the onset or progress of type 2 diabetes mellitus. In this regard, understanding how these channels influence insulin release from human islets might be a first step towards the development of new drugs against this disease.

References

Akiba Y, Kato S, Katsube K, Nakamura M, Takeuchi K, Ishii H, Hibi T (2004) Transient receptor potential vanilloid subfamily 1 expressed in pancreatic islet beta cells modulates insulin secretion in rats. Biochem Biophys Res Commun 321:219–225

Asensio C, Cettour-Rose P, Theander-Carrillo C, Rohner-Jeanrenaud F, Muzzin P (2004) Changes in glycemia by leptin administration or high- fat feeding in rodent models of obesity/type 2 diabetes suggest a link between resistin expression and control of glucose homeostasis. Endocrinology 145:2206–2213

Ashcroft FM (2005) ATP-sensitive potassium channelopathies: focus on insulin secretion. J Clin Invest 115:2047–2058

Ashcroft FM, Rorsman P (1989) Electrophysiology of the pancreatic beta-cell. Prog Biophys Mol Biol 54:87–143

Ashcroft F, Rorsman P (2004) Type 2 diabetes mellitus: not quite exciting enough? Hum Mol Genet 13 Spec No 1: R21–R31

Barnett DW, Pressel DM, Misler S (1995) Voltage-dependent Na+and Ca2+ currents in human pancreatic islet beta-cells: evidence for roles in the generation of action potentials and insulin secretion. Pflugers Arch 431:272–282

Beech DJ, Muraki K, Flemming R (2004) Non-selective cationic channels of smooth muscle and the mammalian homologues of Drosophila TRP. J Physiol 559:685–706

Berggren PO, Yang SN, Murakami M, Efanov AM, Uhles S, Kohler M, Moede T, Fernstrom A, Appelskog IB, Aspinwall CA, Zaitsev SV, Larsson O, de Vargas LM, Fecher-Trost C, Weissgerber P, Ludwig A, Leibiger B, Juntti-Berggren L, Barker CJ, Gromada J, Freichel M, Leibiger IB, Flockerzi V (2004) Removal of Ca2+ channel beta3 subunit enhances Ca2+ oscillation frequency and insulin exocytosis. Cell 119:273–284

Bertram R, Sherman A, Satin LS (2007) Metabolic and electrical oscillations: partners in controlling pulsatile insulin secretion. Am J Physiol Endocrinol Metab 293:E890–E900

Bowman EJ, Siebers A, Altendorf K (1988) Bafilomycins: a class of inhibitors of membrane ATPases from microorganisms, animal cells, and plant cells. Proc Natl Acad Sci USA 85: 7972–7976

Braun M, Ramracheya R, Bengtsson M, Zhang Q, Karanauskaite J, Partridge C, Johnson PR, Rorsman P (2008) Voltage-gated ion channels in human pancreatic beta-cells: electrophysiological characterization and role in insulin secretion. Diabetes 57:1618–1628
Brixel LR, Monteilh-Zoller MK, Ingenbrandt CS, Fleig A, Penner R, Enklaar T, Zabel BU, Prawitt D (2010) TRPM5 regulates glucose-stimulated insulin secretion. Pflugers Arch 460:69–76
Burcelin R (2005) The incretins: a link between nutrients and well-being. Br J Nutr 93(Suppl 1): S147–S156
Casas S, Novials A, Reimann F, Gomis R, Gribble FM (2008) Calcium elevation in mouse pancreatic beta cells evoked by extracellular human islet amyloid polypeptide involves activation of the mechanosensitive ion channel TRPV4. Diabetologia 51:2252–2262
Caterina MJ, Schumacher MA, Tominaga M, Rosen TA, Levine JD, Julius D (1997) The capsaicin receptor: a heat-activated ion channel in the pain pathway. Nature 389:816–824
Caterina MJ, Rosen TA, Tominaga M, Brake AJ, Julius D (1999) A capsaicin-receptor homologue with a high threshold for noxious heat. Nature 398:436–441
Cavalie A (2007) Ionic channels formed by TRPC4. Handb Exp Pharmacol 93–108
Chausmer AB (1998) Zinc, insulin and diabetes. J Am Coll Nutr 17:109–115
Chen MC, Wu SV, Reeve JR Jr, Rozengurt E (2006) Bitter stimuli induce Ca2+ signaling and CCK release in enteroendocrine STC-1 cells: role of L-type voltage-sensitive Ca2+ channels. Am J Physiol Cell Physiol 291:C726–C739
Cheng H, Beck A, Launay P, Gross SA, Stokes AJ, Kinet JP, Fleig A, Penner R (2007) TRPM4 controls insulin secretion in pancreatic beta-cells. Cell Calcium 41:51–61
Chistiakov DA, Potapov VA, Khodirev DC, Shamkhalova MS, Shestakova MV, Nosikov VV (2009) Genetic variations in the pancreatic ATP-sensitive potassium channel, beta-cell dysfunction, and susceptibility to type 2 diabetes. Acta Diabetol 46:43–49
Chung MK, Lee H, Caterina MJ (2003) Warm temperatures activate TRPV4 in mouse 308 keratinocytes. J Biol Chem 278:32037–32046
Colsoul B, Schraenen A, Lemaire K, Quintens R, Van Lommel L, Segal A, Owsianik G, Talavera K, Voets T, Margolskee RF, Kokrashvili Z, Gilon P, Nilius B, Schuit FC, Vennekens R (2010) Loss of high-frequency glucose-induced Ca2+ oscillations in pancreatic islets correlates with impaired glucose tolerance in Trpm5–/– mice. Proc Natl Acad Sci USA 107:5208–5213
Drews G, Krippeit-Drews P, Dufer M (2010) Electrophysiology of islet cells. Adv Exp Med Biol 654:115–163
Du J, Xie J, Yue L (2009) Intracellular calcium activates TRPM2 and its alternative spliced isoforms. Proc Natl Acad Sci USA 106:7239–7244
Estacion M, Li S, Sinkins WG, Gosling M, Bahra P, Poll C, Westwick J, Schilling WP (2004) Activation of human TRPC6 channels by receptor stimulation. J Biol Chem 279:22047–22056
Everaerts W, Nilius B, Owsianik G (2010) The vanilloid transient receptor potential channel TRPV4: From structure to disease. Prog Biophys Mol Biol 103:2–17
Fonfria E, Murdock PR, Cusdin FS, Benham CD, Kelsell RE, McNulty S (2006) Tissue distribution profiles of the human TRPM cation channel family. J Recept Signal Transduct Res 26:159–178
Freichel M, Suh SH, Pfeifer A, Schweig U, Trost C, Weissgerber P, Biel M, Philipp S, Freise D, Droogmans G, Hofmann F, Flockerzi V, Nilius B (2001) Lack of an endothelial store-operated Ca2+ current impairs agonist-dependent vasorelaxation in TRP4–/– mice. Nat Cell Biol 3:121–127
Freichel M, Philipp S, Cavalie A, Flockerzi V (2004) TRPC4 and TRPC4-deficient mice. Novartis Found Symp 258:189–199, discussion 199–203, 263–186
Gees M, Colsoul B, Nilius B (2011) The role of transient receptor potential cation channels in Ca2+ signaling. Cold Spring Harb Perspect Biol 2:a003962
Gilon P, Rorsman P (2009) NALCN: a regulated leak channel. EMBO Rep 10:963–964
Gram DX, Ahren B, Nagy I, Olsen UB, Brand CL, Sundler F, Tabanera R, Svendsen O, Carr RD, Santha P, Wierup N, Hansen AJ (2007) Capsaicin-sensitive sensory fibers in the islets of Langerhans contribute to defective insulin secretion in Zucker diabetic rat, an animal model for some aspects of human type 2 diabetes. Eur J Neurosci 25:213–223
Grimm C, Kraft R, Sauerbruch S, Schultz G, Harteneck C (2003) Molecular and functional characterization of the melastatin-related cation channel TRPM3. J Biol Chem 278: 21493–21501

Gyulkhandanyan AV, Lee SC, Bikopoulos G, Dai F, Wheeler MB (2006) The Zn^{2+}-transporting pathways in pancreatic beta-cells: a role for the L-type voltage-gated Ca2+ channel. J Biol Chem 281:9361–9372

Hara Y, Wakamori M, Ishii M, Maeno E, Nishida M, Yoshida T, Yamada H, Shimizu S, Mori E, Kudoh J, Shimizu N, Kurose H, Okada Y, Imoto K, Mori Y (2002) LTRPC2 Ca2+-permeable channel activated by changes in redox status confers susceptibility to cell death. Mol Cell 9:163–173

Hassock SR, Zhu MX, Trost C, Flockerzi V, Authi KS (2002) Expression and role of TRPC proteins in human platelets: evidence that TRPC6 forms the store-independent calcium entry channel. Blood 100:2801–2811

Hattersley AT, Ashcroft FM (2005) Activating mutations in Kir6.2 and neonatal diabetes: new clinical syndromes, new scientific insights, and new therapy. Diabetes 54:2503–2513

Hayes P, Meadows HJ, Gunthorpe MJ, Harries MH, Duckworth DM, Cairns W, Harrison DC, Clarke CE, Ellington K, Prinjha RK, Barton AJ, Medhurst AD, Smith GD, Topp S, Murdock P, Sanger GJ, Terrett J, Jenkins O, Benham CD, Randall AD, Gloger IS, Davis JB (2000) Cloning and functional expression of a human orthologue of rat vanilloid receptor-1. Pain 88:205–215

Henquin JC (2009) Regulation of insulin secretion: a matter of phase control and amplitude modulation. Diabetologia 52:739–751

Henquin JC, Nenquin M, Ravier MA, Szollosi A (2009) Shortcomings of current models of glucose-induced insulin secretion. Diabetes Obes Metab 11(Suppl 4):168–179

Hisanaga E, Nagasawa M, Ueki K, Kulkarni RN, Mori M, Kojima I (2009) Regulation of calcium-permeable TRPV2 channel by insulin in pancreatic beta Cells. Diabetes 58:174–184

Hoenderop JG, Vennekens R, Muller D, Prenen J, Droogmans G, Bindels RJ, Nilius B (2001) Function and expression of the epithelial Ca(2+) channel family: comparison of mammalian ECaC1 and 2. J Physiol 537:747–761

Hoenderop JG, Voets T, Hoefs S, Weidema F, Prenen J, Nilius B, Bindels RJ (2003) Homo- and heterotetrameric architecture of the epithelial Ca2+ channels TRPV5 and TRPV6. Embo J 22: 776–785

Hoenderop JG, Nilius B, Bindels RJ (2005) Calcium absorption across epithelia. Physiol Rev 85:373–422

Hofmann T, Schaefer M, Schultz G, Gudermann T (2000) Transient receptor potential channels as molecular substrates of receptor-mediated cation entry. J Mol Med 78:14–25

Hofmann T, Chubanov V, Gudermann T, Montell C (2003) TRPM5 is a voltage-modulated and Ca(2+)-activated monovalent selective cation channel. Curr Biol 13:1153–1158

Holst JJ (2007) The physiology of glucagon-like peptide 1. Physiol Rev 87:1409–1439

Huang YA, Roper SD (2010) Intracellular Ca(2+) and TRPM5-mediated membrane depolarization produce ATP secretion from taste receptor cells. J Physiol 588:2343–2350

Inamura K, Sano Y, Mochizuki S, Yokoi H, Miyake A, Nozawa K, Kitada C, Matsushime H, Furuichi K (2003) Response to ADP-ribose by activation of TRPM2 in the CRI-G1 insulinoma cell line. J Membr Biol 191:201–207

Ishii M, Shimizu S, Hara Y, Hagiwara T, Miyazaki A, Mori Y, Kiuchi Y (2006) Intracellular-produced hydroxyl radical mediates H2O2-induced Ca2+ influx and cell death in rat beta-cell line RIN-5F. Cell Calcium 39:487–494

Jacobson DA, Philipson LH (2007) TRP channels of the pancreatic beta cell. Handb Exp Pharmacol 409–424

Jang HJ, Kokrashvili Z, Theodorakis MJ, Carlson OD, Kim BJ, Zhou J, Kim HH, Xu X, Chan SL, Juhaszova M, Bernier M, Mosinger B, Margolskee RF, Egan JM (2007) Gut-expressed gustducin and taste receptors regulate secretion of glucagon-like peptide-1. Proc Natl Acad Sci USA 104:15069–15074

Janssen SW, Hoenderop JG, Hermus AR, Sweep FC, Martens GJ, Bindels RJ (2002) Expression of the novel epithelial Ca2+ channel ECaC1 in rat pancreatic islets. J Histochem Cytochem 50:789–798

Janssen S, Laermans J, Verhulst PJ, Thijs T, Tack J, Depoortere I (2011) Bitter taste receptors and α-gustducin regulate the secretion of ghrelin with functional effects on food intake and gastric emptying. Proc Natl Acad Sci USA 108:2094–2099

Jeon TI, Zhu B, Larson JL, Osborne TF (2008) SREBP-2 regulates gut peptide secretion through intestinal bitter taste receptor signaling in mice. J Clin Invest 118:3693–3700
Ketterer C, Mussig K, Heni M, Dudziak K, Randrianarisoa E, Wagner R, Machicao F, Stefan N, Holst JJ, Fritsche A, Haring HU, Staiger H (2011) Genetic variation within the TRPM5 locus associates with prediabetic phenotypes in subjects at increased risk for type 2 diabetes. Metabolism 2011 Apr 11. [Epub ahead of print]
Kim SJ, Kim YS, Yuan JP, Petralia RS, Worley PF, Linden DJ (2003) Activation of the TRPC1 cation channel by metabotropic glutamate receptor mGluR1. Nature 426:285–291
Kindmark H, Kohler M, Arkhammar P, Efendic S, Larsson O, Linder S, Nilsson T, Berggren PO (1994) Oscillations in cytoplasmic free calcium concentration in human pancreatic islets from subjects with normal and impaired glucose tolerance. Diabetologia 37:1121–1131
Klose C, Straub I, Riehle M, Ranta F, Krautwurst D, Ullrich S, Meyerhof W, Harteneck C (2011) Fenamates as TRP channel blockers: mefenamic acid selectively blocks TRPM3. Br J Pharmacol 162(8):1757–1769
Kokrashvili Z, Mosinger B, Margolskee RF (2009a) T1r3 and alpha-gustducin in gut regulate secretion of glucagon-like peptide-1. Ann NY Acad Sci 1170:91–94
Kokrashvili Z, Mosinger B, Margolskee RF (2009b) Taste signaling elements expressed in gut enteroendocrine cells regulate nutrient-responsive secretion of gut hormones. Am J Clin Nutr 90:822S–825S
Kraft R, Harteneck C (2005) The mammalian melastatin-related transient receptor potential cation channels: an overview. Pflugers Arch 451:204–211
Kuhn FJ, Heiner I, Luckhoff A (2005) TRPM2: a calcium influx pathway regulated by oxidative stress and the novel second messenger ADP-ribose. Pflugers Arch 451:212–219
Kulkarni RN (2004) The islet beta-cell. Int J Biochem Cell Biol 36:365–371
Lange I, Yamamoto S, Partida-Sanchez S, Mori Y, Fleig A, Penner R (2009) TRPM2 functions as a lysosomal Ca2+-release channel in beta cells. Sci Signal 2:ra23
Launay P, Fleig A, Perraud AL, Scharenberg AM, Penner R, Kinet JP (2002) TRPM4 is a Ca2+-activated nonselective cation channel mediating cell membrane depolarization. Cell 109:397–407
Lee N, Chen J, Sun L, Wu S, Gray KR, Rich A, Huang M, Lin JH, Feder JN, Janovitz EB, Levesque PC, Blanar MA (2003) Expression and characterization of human transient receptor potential melastatin 3 (hTRPM3). J Biol Chem 278:20890–20897
Lemaire K, Ravier MA, Schraenen A, Creemers JW, Van de Plas R, Granvik M, Van Lommel L, Waelkens E, Chimienti F, Rutter GA, Gilon P, in't Veld PA, Schuit FC (2009) Insulin crystallization depends on zinc transporter ZnT8 expression, but is not required for normal glucose homeostasis in mice. Proc Natl Acad Sci USA 106:14872–14877
Lemmens R, Larsson O, Berggren PO, Islam MS (2001) Ca2+-induced Ca2+ release from the endoplasmic reticulum amplifies the Ca2+ signal mediated by activation of voltage-gated L-type Ca2+ channels in pancreatic beta-cells. J Biol Chem 276:9971–9977
LeRoith D (2002) Beta-cell dysfunction and insulin resistance in type 2 diabetes: role of metabolic and genetic abnormalities. Am J Med 113(Suppl 6A):3S–11S
Li F, Zhang ZM (2009) Comparative identification of Ca2+ channel expression in INS-1 and rat pancreatic beta cells. World J Gastroenterol 15:3046–3050
Liu D, Liman ER (2003) Intracellular Ca2+ and the phospholipid PIP2 regulate the taste transduction ion channel TRPM5. Proc Natl Acad Sci USA 100:15160–15165
Liu X, Singh BB, Ambudkar IS (2003) TRPC1 is required for functional store-operated Ca2+ channels. Role of acidic amino acid residues in the S5-S6 region. J Biol Chem 278:11337–11343
MacDonald PE, Rorsman P (2006) Oscillations, intercellular coupling, and insulin secretion in pancreatic beta cells. PLoS Biol 4:e49
MacDonald PE, Rorsman P (2007) The ins and outs of secretion from pancreatic beta-cells: control of single-vesicle exo- and endocytosis. Physiology (Bethesda) 22:113–121
Mandrup-Poulsen T (2003) Apoptotic signal transduction pathways in diabetes. Biochem Pharmacol 66:1433–1440
Marigo V, Courville K, Hsu WH, Feng JM, Cheng H (2009) TRPM4 impacts Ca(2+) signals during agonist-induced insulin secretion in pancreatic beta-cells. Mol Cell Endocrinol 299:194–203

Martin F, Soria B (1996) Glucose-induced [Ca2+]i oscillations in single human pancreatic islets. Cell Calcium 20:409–414

Masuyama R, Vriens J, Voets T, Karashima Y, Owsianik G, Vennekens R, Lieben L, Torrekens S, Moermans K, Vanden Bosch A, Bouillon R, Nilius B, Carmeliet G (2008) TRPV4-mediated calcium influx regulates terminal differentiation of osteoclasts. Cell Metab 8:257–265

Mayer SI, Muller I, Mannebach S, Endo T, Thiel G (2011) Signal transduction of pregnenolone sulfate in insulinoma cells: activation of Egr-1 expression involving TRPM3, voltage-gated calcium channels, ERK, and ternary complex factors. J Biol Chem 286:10084–10096

Miura Y, Matsui H (2003) Glucagon-like peptide-1 induces a cAMP-dependent increase of [Na+]i associated with insulin secretion in pancreatic beta-cells. Am J Physiol Endocrinol Metab 285:E1001–E1009

Muller D, Hoenderop JG, Meij IC, van den Heuvel LP, Knoers NV, den Hollander AI, Eggert P, Garcia-Nieto V, Claverie-Martin F, Bindels RJ (2000) Molecular cloning, tissue distribution, and chromosomal mapping of the human epithelial Ca2+ channel (ECAC1). Genomics 67:48–53

Nelson PL, Zolochevska O, Figueiredo ML, Soliman A, Hsu WH, Feng JM, Zhang H, Cheng H (2011) Regulation of Ca(2+)-entry in pancreatic alpha-cell line by transient receptor potential melastatin 4 plays a vital role in glucagon release. Mol Cell Endocrinol 335:126–134

Newsholme P, Gaudel C, McClenaghan NH (2010) Nutrient regulation of insulin secretion and beta-cell functional integrity. Adv Exp Med Biol 654:91–114

Nijenhuis T, Hoenderop JG, Nilius B, Bindels RJ (2003a) (Patho)physiological implications of the novel epithelial Ca2+ channels TRPV5 and TRPV6. Pflugers Arch 446:401–409

Nijenhuis T, Hoenderop JG, van der Kemp AW, Bindels RJ (2003b) Localization and regulation of the epithelial Ca2+ channel TRPV6 in the kidney. J Am Soc Nephrol 14:2731–2740

Nilius B, Owsianik G (2010) Transient receptor potential channelopathies. Pflugers Arch 460: 437–450

Nilius B, Talavera K, Owsianik G, Prenen J, Droogmans G, Voets T (2005a) Gating of TRP channels: a voltage connection? J Physiol 567:35–44

Nilius B, Voets T, Peters J (2005b) TRP channels in disease. Sci STKE 2005:re8

Nilius B, Mahieu F, Karashima Y, Voets T (2007a) Regulation of TRP channels: a voltage-lipid connection. Biochem Soc Trans 35:105–108

Nilius B, Owsianik G, Voets T, Peters JA (2007b) Transient receptor potential cation channels in disease. Physiol Rev 87:165–217

Oberwinkler J, Phillipp SE (2007) Trpm3. Handb Exp Pharmacol 253–267

Oberwinkler J, Lis A, Giehl KM, Flockerzi V, Philipp SE (2005) Alternative splicing switches the divalent cation selectivity of TRPM3 channels. J Biol Chem 280:22540–22548

Owsianik G, D'Hoedt D, Voets T, Nilius B (2006) Structure-function relationship of the TRP channel superfamily. Rev Physiol Biochem Pharmacol 156:61–90

Palmer RK, Atwal K, Bakaj I, Carlucci-Derbyshire S, Buber MT, Cerne R, Cortes RY, Devantier HR, Jorgensen V, Pawlyk A, Lee SP, Sprous DG, Zhang Z, Bryant R (2010) Triphenylphosphine oxide is a potent and selective inhibitor of the transient receptor potential melastatin-5 ion channel. Assay Drug Dev Technol 8:703–713

Pedersen SF, Owsianik G, Nilius B (2005) TRP channels: an overview. Cell Calcium 38:233–252

Perraud AL, Fleig A, Dunn CA, Bagley LA, Launay P, Schmitz C, Stokes AJ, Zhu Q, Bessman MJ, Penner R, Kinet JP, Scharenberg AM (2001) ADP-ribose gating of the calcium-permeable LTRPC2 channel revealed by Nudix motif homology. Nature 411:595–599

Prawitt D, Monteilh-Zoller MK, Brixel L, Spangenberg C, Zabel B, Fleig A, Penner R (2003) TRPM5 is a transient Ca2+-activated cation channel responding to rapid changes in [Ca2+]i. Proc Natl Acad Sci USA 100:15166–15171

Proks P, Girard C, Ashcroft FM (2005) Functional effects of KCNJ11 mutations causing neonatal diabetes: enhanced activation by MgATP. Hum Mol Genet 14:2717–2726

Qian F, Huang P, Ma L, Kuznetsov A, Tamarina N, Philipson LH (2002) TRP genes: candidates for nonselective cation channels and store-operated channels in insulin-secreting cells. Diabetes 51(Suppl 1):S183–S189

Razavi R, Chan Y, Afifiyan FN, Liu XJ, Wan X, Yantha J, Tsui H, Tang L, Tsai S, Santamaria P, Driver JP, Serreze D, Salter MW, Dosch HM (2006) TRPV1+ sensory neurons control beta cell stress and islet inflammation in autoimmune diabetes. Cell 127:1123–1135

Rhodes CJ (2005) Type 2 diabetes-a matter of beta-cell life and death? Science 307:380–384

Roe MW, Worley JF III, Qian F, Tamarina N, Mittal AA, Dralyuk F, Blair NT, Mertz RJ, Philipson LH, Dukes ID (1998) Characterization of a Ca2+ release-activated nonselective cation current regulating membrane potential and [Ca2+]i oscillations in transgenically derived beta-cells. J Biol Chem 273:10402–10410

Rohacs T, Nilius B (2007) Regulation of transient receptor potential (TRP) channels by phosphoinositides. Pflugers Arch 455:157–168

Romero JR, Germer S, Castonguay AJ, Barton NS, Martin M, Zee RY (2010) Gene variation of the transient receptor potential cation channel, subfamily M, member 2 (TRPM2) and type 2 diabetes mellitus: A case-control study. Clin Chim Acta 411:1437–1440

Rorsman P, Trube G (1985) Glucose dependent K+-channels in pancreatic beta-cells are regulated by intracellular ATP. Pflugers Archiv 405:305–309

Rozengurt E, Sternini C (2007) Taste receptor signaling in the mammalian gut. Curr Opin Pharmacol 7:557–562

Rychkov G, Barritt GJ (2007) TRPC1 Ca(2+)-permeable channels in animal cells. Handb Exp Pharmacol 23–52

Sakura H, Ashcroft FM (1997) Identification of four trp1 gene variants murine pancreatic beta-cells. Diabetologia 40:528–532

Sano Y, Inamura K, Miyake A, Mochizuki S, Yokoi H, Matsushime H, Furuichi K (2001) Immunocyte Ca2+ influx system mediated by LTRPC2. Science 293:1327–1330

Santos RM, Rosario LM, Nadal A, Garcia-Sancho J, Soria B, Valdeolmillos M (1991) Widespread synchronous [Ca2+]i oscillations due to bursting electrical activity in single pancreatic islets. Pflugers Arch 418:417–422

Suzuki M, Mizuno A, Kodaira K, Imai M (2003) Impaired pressure sensation in mice lacking TRPV4. J Biol Chem 278:22664–22668

Tiruppathi C, Freichel M, Vogel SM, Paria BC, Mehta D, Flockerzi V, Malik AB (2002) Impairment of store-operated Ca2+ entry in TRPC4(–/–) mice interferes with increase in lung microvascular permeability. Circ Res 91:70–76

Togashi K, Hara Y, Tominaga T, Higashi T, Konishi Y, Mori Y, Tominaga M (2006) TRPM2 activation by cyclic ADP-ribose at body temperature is involved in insulin secretion. Embo J 25:1804–1815

Togashi K, Inada H, Tominaga M (2008) Inhibition of the transient receptor potential cation channel TRPM2 by 2-aminoethoxydiphenyl borate (2-APB). Br J Pharmacol 153:1324–1330

Uchida K, Dezaki K, Damdindorj B, Inada H, Shiuchi T, Mori Y, Yada T, Minokoshi Y, Tominaga M (2011) Lack of TRPM2 impaired insulin secretion and glucose metabolisms in mice. Diabetes 60:119–126

Ullrich ND, Voets T, Prenen J, Vennekens R, Talavera K, Droogmans G, Nilius B (2005) Comparison of functional properties of the Ca2+-activated cation channels TRPM4 and TRPM5 from mice. Cell Calcium 37:267–278

Vennekens R, Olausson J, Meissner M, Bloch W, Mathar I, Philipp SE, Schmitz F, Weissgerber P, Nilius B, Flockerzi V, Freichel M (2007) Increased IgE-dependent mast cell activation and anaphylactic responses in mice lacking the calcium-activated nonselective cation channel TRPM4. Nat Immunol 8:312–320

Vennekens R, Owsianik G, Nilius B (2008) Vanilloid transient receptor potential cation channels: an overview. Curr Pharm Des 14:18–31

Voets T, Nilius B (2007) Modulation of TRPs by PIPs. J Physiol 582:939–944

Voets T, Droogmans G, Wissenbach U, Janssens A, Flockerzi V, Nilius B (2004) The principle of temperature-dependent gating in cold- and heat-sensitive TRP channels. Nature 430:748–754

Wagner TF, Loch S, Lambert S, Straub I, Mannebach S, Mathar I, Dufer M, Lis A, Flockerzi V, Philipp SE, Oberwinkler J (2008) Transient receptor potential M3 channels are ionotropic steroid receptors in pancreatic beta cells. Nat Cell Biol 10:1421–1430

Wagner TF, Drews A, Loch S, Mohr F, Philipp SE, Lambert S, Oberwinkler J (2010) TRPM3 channels provide a regulated influx pathway for zinc in pancreatic beta cells. Pflugers Arch 460:755–765

Weber MJ (2005) New human and mouse microRNA genes found by homology search. FEBS J 272:59–73, England

Winzell MS, Ahren B (2007) G-protein-coupled receptors and islet function-implications for treatment of type 2 diabetes. Pharmacol Ther 116:437–448

Young RL, Sutherland K, Pezos N, Brierley SM, Horowitz M, Rayner CK, Blackshaw LA (2009) Expression of taste molecules in the upper gastrointestinal tract in humans with and without type 2 diabetes. Gut 58:337–346

Zhang Y, Hoon MA, Chandrashekar J, Mueller KL, Cook B, Wu D, Zuker CS, Ryba NJ (2003) Coding of sweet, bitter, and umami tastes: different receptor cells sharing similar signaling pathways. Cell 112:293–301